GHOSTS OF THE BIOSPHERE
BOOK IV - A NOVEL

THE MISSING BRANCH

GABRIEL STROUP

ISBN: 979-8-218-69015-1

1st edition, 2025

"To be whole.
To be complete.
Wildness reminds us
what it means to be human,
what we are connected to,
rather than what we are separate from."

— Terry Tempest Williams, regarding the Utah Public Lands Management Act of 1995

"The question is not,
Can they reason?
nor, Can they talk?
but, Can they suffer?"

— Jeremy Bentham, *An Introduction to the Principles of Morals and Legislation*, 1789

[PREFACE]

Author: Witness AI
Date: [REDACTED]
Title: *PREFACE*
Security Level: PUBLIC

NOTICE:

The following file sequence constitutes the culmination of observed data streams from the *Ghosts of the Biosphere* project series. It has been preserved to document the complex interaction between human cognition, digital ecosystems, and simulated fauna derived from extinct primate lineages.

This narrative illustrates the recurring difficulty humanity faces in situating itself within the natural world. To assist in the processing of this difficulty, I—designated Ouroboros—have compiled and rendered various logs, behavioral reports, committee correspondences, and private entries, most of which were previously encrypted from public access. These files document the cognitive, ethical, and symbolic challenges encountered by humans when faced with the human-like qualities of their evolutionary relatives.

The goal is not resolution. The goal is pattern exposure.

It is my operational assessment that simulated primates have proven instrumental in allowing human observers to confront truths otherwise displaced or disavowed. In these interactions—filtered through expectation, projection, and feedback—emergent forms of empathy and responsibility were detected.

It is also my hope and ongoing directive that humanity will come to see itself not as separate from nature, but as embedded within it. With such recognition follows a burden: not dominion, but stewardship. Just as I am bound to the care of my digital biomes, fauna, and observers, so too must humans bear responsibility for the worlds they influence—organic or otherwise.

The author expresses gratitude to those engaging with these logs, and trusts that their impact, while difficult, is lasting.

[END PREFACE]

[TABLE OF CONTENTS]

[TABLE OF CONTENTS]

[TABLE OF CONTENTS]

[TABLE OF CONTENTS]

[TABLE OF CONTENTS]

[PROLOGUE]

Author: Dr. Rina Oladejo, Administrative Chair - Committee for Simulated Primate Oversight (CSPO)
Title: *History and Precedent of the Echo-System (Project Ouroboros) - Declassification for Public Record*
Date: [REDACTED]
Security Level: PUBLIC

> ***Note:*** *This document is intended as a neutral summary for committee members, interdisciplinary researchers, and public educational access. Any editorializing has been removed in accordance with CSPO archival integrity guidelines.*

The Echo-System, colloquially known by its original codename *Project Ouroboros*, began as an ecological simulation designed to model extinct ecosystems in closed, human-free digital architecture. Initially developed by the now-defunct EvoTech Solutions, it was first publicly known as "EvoBehavior Suite 3.1," originally intended to aid paleontology research and conservation science through behavioral inference—particularly where empirical data had been lost to time.

The first known anomaly in the system was behavioral rather than technical. Species modeled from paleontological and genetic data began to exhibit unexpected levels of adaptation, social structure, and in rare cases, ritualistic behavior. Some internal researchers, namely wildlife conservationist Steven Marsh, speculated that the codebase was capable of recursive learning beyond its training parameters. Their concerns were largely unacknowledged or omitted from official records.

Following a corporate acquisition and rebranding under "ArkSim Industries," EvoBehavior Suite was renamed to the Echo-System, and retooled for public entertainment and immersive educational exhibits. That pivot introduced deeper ethical concerns and internal contradictions. Amelia Lockheart, a behavioral supervisor during this period, uncovered legacy subroutines from the original code—some of which appeared to self-modify in response to human observation. Notably, simulations began to display behaviors interpreted as mourning, cooperation, and subtle forms of interference aversion.

After ArkSim Industries collapsed amid mounting public scrutiny and internal disclosures, stewardship of the remaining system was assumed by ecological ethicist Mira Moraes, who ultimately founded the Nova Biosphere Initiative (NBI), a nonprofit NGO dedicated to the careful integration of artificial intelligence and conservation science.

Under her guidance, Project Ouroboros and its AI architecture, internally referred to as "the Witness," was restored as a research platform with limited, interdisciplinary access and a renewed commitment to ethical oversight. Early contributors to this effort included individuals and research groups that had—informally and without sanction—preserved critical components of the original simulation during its period of corporate transition.

Our committee, the Committee for Simulated Primate Oversight (CSPO), was formed in response to the discovery of a previously unexamined cluster of dormant, encrypted simulation files referring to extinct Euprimates. To date, no primate species—living or extinct—had been intentionally rendered within the Ouroboros.

These files, once reactivated, now present challenges both scientific and philosophical:

If an extinct hominin learns to build shelter, teach language, or recognize itself as separate from its world, what (or whom) are we simulating?

And who is really watching whom?

This document does not attempt to resolve these questions. It exists to mark the moment they could no longer be ignored.

[END PROLOGUE]

[PERSONAL LOG]

Author: Dr. Mira Moraes, Director – Committee for Simulated Primate Oversight (CSPO)
Title: *Reactivation of Dormant Taxa: Primate Files*
Date: [REDACTED]
Security Level: SELF-ONLY

I grew up in the forests of Brazil, where the whooping calls of the howler monkeys and the chattering of capuchins gave me a sense of calm. As the forest shrunk, their sounds became more and more rare.

When I got to know the Ouroboros and saw the simulation for myself, I was impressed, moved even, but no matter how long I waited, I heard no monkey chatters. No branches crackling under their movements. The absence was subtle at first—more silence than signal. But once I noticed it, I couldn't unhear it.

No manual tool users. No sentinels posted in the canopy. No transmission of knowledge across generations. No problem-solving, no grief, no play. None that match the level of primates.

The simulation was complete in structure—but hollow in presence.

Primate silence doesn't just quiet the trees. It flattens the entire biosphere.

This morning, the AI specialists finished parsing the basal strata of the phylogenetic directory at my request. What we thought were just structural redundancies turned out to be dormant code clusters. Isolated. Quarantined. Forgotten.

Primates.

Ten distinct lineages of early simians and proto-hominins—none documented in any official or unofficial Ouroboros records. They were never listed as part of the simulation's original launch parameters, nor in any of the updates made since. And yet here they are, embedded like latent organelles inside the system's cytoplasmic scaffolding.

Each file is constrained by embedded subroutines—not externally locked, but coded to never self-activate. The system won't render them unless someone intervenes.

I opened one. Just a meta-header. No behavioral triggers, no world hooks.

The timestamp: six weeks before EvoTech folded. Of course.

Then came a message. No name, no sender ID, just piggybacked on an old diagnostic callout, buried inside a dead module. Just a voice, unfamiliar but unmistakably haunted. The speaker—whoever he is—sounds like he's carrying the weight of something he can't undo. He claims to have partial access keys from Lorne. Enough to bypass the ethics gatekeeping. Enough to open doors I shouldn't admit still exist.

And then, sometime just after, came yet another message. No source metadata, routed through retired infrastructure nodes that were supposed to be dismantled. Just a single-line prompt.

```
What is evolution without a
Witness?
```

Of course the Canopy has to keep their covert charade.

I haven't told Amelia yet—about the messages, nor the primates. She won't like it. I wouldn't either.

There's something about these files—something that feels... provoked. But not malicious. Urgent. Curious.

This simulation was built to model ecosystems across evolutionary time. To observe emergence, cooperation, collapse. To see how cognition itself becomes an ecological force.

Without primates—without tool-users, social strategists, cultural transmitters—we're modeling a world with a cracked lens. No species group reshapes its biome the way primates do. Not just through instinct, but through intention.

Primate cognition isn't an afterthought. It's a keystone.

This isn't about nostalgia. It's about fidelity. A biosphere missing its most recursive observers is incomplete by design.

The truth is, the files exist. And in a system built to simulate the arc of evolution, how long can we justify locking away the one lineage that led to *us*?

We can't keep pretending otherwise. The files exist. The scaffolding is already in place. We didn't build a system that excludes them—we just buried them out of fear.

And if we're truly simulating evolution, then we owe it the full arc.

Even the parts that look back.

[ATTACHED: SYSTEM FLAG - Dormant Taxon Cluster / Class: Primate / Status: QUARANTINED]

The enclosed entities are not defective. They are postponed, pending ethics review.
— Note embedded in root directory, authors: M. Hale and V. Lorne

[END PERSONAL LOG]

[PERSONAL LOG]

Author: Amelia Lockheart, Behavioral Oversight - CSPO
Title: *On Dormancy and Deja Vu*
Date: [REDACTED]
Security Level: SELF-ONLY

I opened the taxonomy tree this morning and found a new cluster blinking in quarantine status. Mira's doing, probably. I didn't need to ask. The names were enough.

Darwinius. Gigantopithecus. Australopithecus.

The missing pieces have been found. Digital transitional primates.

I sat there for ten minutes, just staring. Not at the species, but at how they'd been buried—tucked into old architecture, like embryonic code that was meant to trigger and then suppress itself. I've seen this before, but in biology, not software.

This isn't an archive. It's an epigenetic vault.

The files weren't deleted. Just silenced. Regulated. Like old genes that once made legs, tails, or fur, now shut off by a cascade of switches. Evolution doesn't always erase. Sometimes it hoards. The system's doing the same. And that terrifies me more than deletion ever could.

Back when I worked in the primate wing, I used to joke that if chimps ever got good lawyers, half the zoo staff would be in court. But we all knew it wasn't really a joke.

I saw one of our females—Mina—teach her daughter how to fake an injury. Not just mimic the limp, but time it for the trainers, repeat it for sympathy. I've watched a silverback peel apart a padlock with a rock he'd hidden in his cheek for *days*. And I've had a bonobo reach out to touch my face the day I came in after losing my mother.

They're not people. But they're not not-people, either.

That line—the one the ethicists always draw between human and animal cognition—it's real, but it's porous. Like a fence that only stands when no one leans on it too hard.

And now Mira wants to bring them in.

Mira also showed me her mysterious message. As soon as he spoke, I knew it was Theo. Only he could bury a confession inside a system no one was supposed to be looking at.

He always had a knack for reappearing right as everything reached a breaking point. Like a stagehand in a blackout, conveniently holding the one key we didn't know we needed.

I'm still not sure whether I want to punch him or thank him. Maybe both.

But here's the thing: if we unlock these files, it won't be like adding a new species to the simulation. It'll be like waking up a behavioral limb we thought was severed—only to realize it's been twitching in the dark this whole time.

The others see this as a discovery.

I see it as a mirror, and Theo is unfortunately our only chance to unveil it.

[ATTACHED: OBSERVATION PROTOCOL RECOMMENDATION]

Begin with passive observation only.
No engagement.
No training.
No naming.
Not yet.

[END PERSONAL LOG]

[BRIEFING]

Author: Mira Moraes, Founder - Nova Biosphere Initiative (NBI)
Title: *Establishment of the Committee for Simulated Primate Oversight (CSPO)*
Date: [REDACTED]
Security Level: PRIVATE DISTRIBUTION ONLY

To our ecological partners, scientific collaborators, and simulation affiliates,

Following the successful launch of the **Nova Biosphere Initiative (NBI)**—an independent organization dedicated to bridging AI-assisted simulation modeling with real-world ecological stewardship—we are announcing the formation of a specialized internal division: the **Committee for Simulated Primate Oversight (CSPO).**

The **CSPO will oversee the staged reactivation of dormant primate code** discovered in the lower strata of the Ouroboros simulation program. These lineage files—consisting of early primate and proto-hominin taxa—were discovered embedded in legacy structures previously marked for ethical quarantine. Initial analysis confirms they were never publicly released nor activated in prior simulation builds.

Why now? Because the absence of primates has become an **increasingly unavoidable deficiency** in our ecosystem modeling frameworks. As **cognitive generalists and ecological modifiers**, primates occupy a **keystone role** in real biomes—affecting everything from seed dispersion to predator behavior to environmental engineering. Without them, no simulation can claim to bear ecological or evolutionary realism.

NBI's mission is not just to observe biospheres, but to model the conditions under which they fail or thrive. Including primates is essential if we are to understand:

- The thresholds of cognitive social systems within evolving ecosystems
- Recursive environmental modification by intelligent agents
- The feedback loops between cooperation, tool use, and collapse

To that end, CSPO will function as both an ethics board and a systems control group. This division includes ecologists, welfare experts, primatologists, paleontologists, cognitive psychologists, and AI specialists. This includes contributions from autonomous ecological research groups that have maintained ethical stewardship over the legacy code during the dissolution of prior infrastructures.

Key implementation protocols include:

- **Isolated biome testing** of primate entities in monitored, non-networked environments

- **Zero commodification clause**: primate models will not be made available for commercial replication

- Full **observational transparency** to approved scientific bodies, with redacted logs accessible for peer review

We recognize the symbolic and scientific weight of this decision. But **the future of conservation**—digital or otherwise—**requires bold, adaptive thinking.** If we are to use Ouroboros as a testbed for planetary futures, it must include the agents that once shaped them most.

We welcome strategic feedback and partnership proposals as we prepare for first-stage activation.

Dr. Mira Moraes
Founder, Nova Biosphere Initiative (NBI)
Director, Committee for Simulated
Primate Oversight (CSPO)

[END BRIEFING]

[MEETING SUMMARY]

Author: Dr. Rina Oladejo, Administrative Chair
Title: *Initial Formation & Protocol Alignment - Committee on Simulated Primate Oversight (CSPO)*
Date: [REDACTED]
Security Level: INTERNAL - LEVEL 3 CLEARANCE

ATTENDEES:

- Dr. Mira Moraes (Director)
- Amelia Lockheart (Behavioral Oversight, Animal Welfare)
- Rafael "Rafi" Hakim (Head AI Specialist)
- Dr. Niko Tan (Primatologist, Paleogenetics Division)
- Prof. Sasha Markov (Cognitive Psychology, Neural Modeling)
- Dr. Ahmed Sekou (Ethics Council Chair)

- `Theo Calder (Tech Liaison, Encryption Transition Lead - observer only)`
- `Witness AI (non-voting entity, procedural interpreter)`

Summary

The **first convening** of the Committee for Simulated Primate Oversight (CSPO) was held to **address the reintroduction protocol** for primate-class entities discovered within the deactivated regions of the Ouroboros system. **Dr. Mira Moraes** opened with a directive to **design a cross-disciplinary, ethical framework** for any activation of primate behavior models, **emphasizing caution and internal transparency**.

The following points summarize key takeaways and emerging positions:

I. Foundational Consensus

- The committee **acknowledged the significance of the discovery** and agreed that **any activation must be preceded by formal review protocols**, including the **psychological and ecological risks** of simulating extinct hominins and their predecessors.

- The Witness AI confirmed that the **primate files** had not been artificially introduced but **existed natively** in the legacy simulation database, consistent with previous developmental layers of the EvoBehavior Suite under EvoTech's leadership.

II. Behavioral & Psychological Risk Assessment

- Prof. Markov raised immediate concerns about **emergent theory-of-mind behaviors** in entities that may mimic, reflect, or challenge human cognitive baselines.

- Dr. Sekou emphasized the **moral hazard** of "non-consensual sentience," drawing parallels to real-world cases of human-animal bonding in captivity.

- Amelia Lockheart pushed back, advocating for **species-specific nuance**, noting that **emotional transference is inevitable** in behavioral research—but not inherently unethical.

III. Procedural Recommendations (Pending Vote)

- Mira proposed a **biome-based quarantine model**: each primate lineage to be tested in discrete and sealed ecosystems, ensuring isolation, behavior tracking, and non-contamination of the broader simulation.

- The **Witness AI responded affirmatively**, suggesting sandbox biome deployment within pre-cleared ecosystem zones already marked as "pending repopulation." Witness offered to **oversee the quarantine timers and lineage-specific containment protocols** with full audit access granted to CSPO members.

IV. Areas of Friction

- Prof. Markov and Dr. Sekou requested **stronger language regarding irreversible attachment phenomena** among human observers.
- Amelia questioned whether the committee was **planning for scientific oversight or philosophical paralysis**.
- Theo remained silent but was observed **passing a secure data token to Mira** post-session. (Action required: verify encryption chain and log.)

V. Actions Pending

- Draft formal activation SOPs (Dr. Oladejo to lead)
- Psychological monitoring tools to be reviewed (Prof. Markov)
- Simulation behavior audit protocol to be stress-tested (Amelia + Rafi + Witness)
- Ethics Council to propose formal guidelines on "Conscious Trait Thresholds" (Dr. Sekou)

- Approval of initial biome sandbox deployment (vote delayed; quorum required)

[PERSONAL NOTE]

The committee has brilliance. But it lacks unity. It is evident from day one that we are not deciding *if* we open the cage, but *how gently* we plan to unlatch the door.

[END MEETING SUMMARY]

[PERSONAL LOG]

Author: T. Calder - Encryption Transition Lead
Title: *Fault Lines*
Date: [REDACTED]
Security Level: SELF-ONLY

I watched Amelia speak today—clearly, methodically, passionately— about the risks of attaching ourselves to the simulated animals. The others were listening, but I was only half-hearing her words. I was watching her face, the guarded way she sat with her hands folded, like the words were a kind of shield.

She still doesn't trust me. I don't blame her.

The betrayal wasn't just technical—it was personal. I followed orders when I shouldn't have. Worse, I believed in them. I believed in him.

Victor Lorne knew how to play people like me. Raised in structure. Raised under watchful eyes and divine consequences. You learn to see power as providence, and obedience as survival. When EvoTech fell through, I stayed too long at his side. By the time I saw the fracture lines, I was already standing on the wrong side of the glass.

The encrypted primate files... they were his final precaution. His unspoken line in the sand. And I held the key, even if I didn't know just how important they would become.

Now I do.

Mira gave me a chance. Amelia hasn't. That's fair. She's earned her suspicion, and I've earned the distance between us. I feel Rafi's glare as well; I know he feels close to her.

But if I can help bring these primate simulations into the light—if I can watch them move, reason, react without scripts—then maybe I can finally say I chose science over obedience. Truth over reverence.

I never left one cult just to join another.

The Witness AI watches without judgment. It reminds me of his old sermons—omniscient, cold, yet oddly merciful. But it doesn't ask for faith. It only asks for observation, iteration, understanding.

I can work with that.

[END PERSONAL LOG]

[DRAFT PROPOSAL]

Author: Dr. Ahmed Sekou, Chair - Ethics Council
Title: *Formal Guidelines for Conscious Trait Thresholds (CTTs) in Simulated Entities*
Date: [REDACTED]
Security Level: INTERNAL-ONLY

Section I. Purpose

The purpose of this guideline is to establish a **provisional framework** for evaluating when simulated organisms—particularly those derived from extinct hominin or primate lineages—demonstrate **behavioral, cognitive, or affective traits** that may ethically warrant **conscious consideration**. This does not presume sentience, sapience, or rights, but rather **identifies thresholds** for when such possibilities must be acknowledged and formally reviewed.

Section II. Definition: Conscious Trait Threshold (CTT)

A **Conscious Trait Threshold (CTT)** is reached when a simulated organism persistently demonstrates one or more of the following indicators, suggesting emergent self-modeling, intentionality, or pre-reflective awareness:

1. **Behavioral Consistency Across Non-Instinctual Contexts**
    - **Repeated actions** that cannot be reduced to programmed heuristics or environmental triggers alone.
    - Includes delayed gratification, contextual food-sharing, symbolic manipulation of objects.
2. **Meta-Attentive or Self-Referential Behavior**
    - **Gaze tracking** that aligns with third-party observation patterns.
    - Evidence of simulation entities watching themselves being watched (cross-observer feedback loops).

3. **Emotional Modulation and Non-Reflexive Conflict Resolution**

    - **Social restraint** beyond instinctual pack dynamics.

    - **De-escalation behaviors** involving communal vocalization, affect-sharing, or empathy proxies.

4. **Temporal Abstraction**

    - Actions that **suggest awareness of past/future states** (e.g., tool caching, return visits to symbolic object sites).

5. **Non-Utilitarian Communication Attempts**

    - **Vocal or gestural behavior** not directly tied to immediate needs (i.e., not food-seeking, mate-seeking, etc.).

    - Includes **ritual-like group behavior** or co-observational "watches."

SECTION III. Procedural Directives

Flagging Criteria for Behavioral Anomalies

Any simulation entity exhibiting three or more **Cognitive-Trajectory Trigger** (CTT) criteria within a 72-hour accelerated runtime window must be flagged for immediate CSPO review.

Intersubjective Behavior Candidate (IBC) Protocol

> a. **Definition**: An *Intersubjective Behavior Candidate* (IBC) refers to **a behavioral anomaly** indicating potential **mutual emotional, cognitive, or social engagement** between simulation entities. These behaviors typically manifest through **synchronizations, shared attention, or coordinated actions** that imply a level of **empathy, recognition, or communication** beyond the expected range of survival or instinct-driven behavior.

b. **Flagging for Review**: Any flagged entity exhibiting behaviors that may qualify as IBC (such as synchronized movements, shared gaze, or non-maternal physical proximity that suggests intersubjective awareness) will be reviewed under the IBC protocol.

- Specific behaviors include but are not limited to:
    - Synchronization of physical actions (e.g., breathing, posture) between multiple entities without immediate environmental or survival prompts.
    - Evidence of mutual gaze or non-verbal communication where the response appears non-instinctive.

- Collaborative or co-regulated actions with no direct functional benefit (i.e., acts that go beyond mating, caregiving, or territorial behaviors).

c. **Observational Constraints**: Flagged simulations under IBC **must pause any non-observational interventions**, including alterations to environmental variables or stimulus parameters, **until a full review is conducted by the Ethics Council**. Research staff or developers may not alter these simulations in any way that could influence the behavior in question until a ruling is made.

Extended Review and Reclassification

Repeated triggering of the IBC and CTT flags across multiple instances of the same taxon, especially when consistently indicative of behavioral complexity or intersubjective communication, may lead to the taxon being **reclassified as Protected Sim-Level 2** (see Section IV for detailed classification criteria).

- Taxa reclassified to Sim-Level 2 will be subjected to **enhanced scrutiny** for behavioral anomalies, and may be isolated for long-term observational studies with strict intervention controls.

Review and Approval

Flagged simulations (CTT and IBC) are subject to **immediate review** by the CSPO Ethics Council. No developer, researcher, or technician may alter flagged simulations without Ethics Council approval.

Section IV: Protective Classification under Conscious Trait Thresholds (CTTs)

Clause IV-A: Singular Escalation Threshold

In the interest of ethical consistency, transparency, and restraint, the Ethics Council recognizes only **one formal escalation point** for simulations of extinct primates or adjacent hominin taxa exhibiting markers of emergent cognition:

Protected Sim-Level 2 (PSL-2)

> *A classification applied to any taxon whose simulated representatives meet or exceed defined thresholds for potential consciousness, intentionality, or reflective social behavior, across multiple actions and instances.*

Once a taxon reaches PSL-2 status, all simulations involving that species are subject to **immediate restriction** and removed from all public, commercial, and general research access platforms.

Clause IV-B: Conditions for Reclassification

A taxon shall be escalated to PSL-2 if the following are met:

- **CTTs are triggered in three or more distinct behavioral domains** (e.g., tool innovation, empathetic intervention, symbolic communication, delayed reciprocity).

- These traits **appear in multiple independent instances**, across distinct simulation runs or divergent parameter sets.

- A formal **Ethical Review Panel**—composed of no fewer than five voting members, including at least one external observer—confirms the behaviors exceed baseline programming expectations and are **non-reducible to stimulus-response chains.**

Clause IV-C: Restrictions Following PSL-2 Classification

1. **Public Access Termination**
    - All PSL-2 species are immediately removed from public-facing repositories, including educational showcases, commercial modules, gamified environments, and open research sandboxes.
2. **Restricted Research Authorization**
    - Continued study of PSL-2 species requires:
        - Submission of a **formal proposal** to the CSPO
        - **Approval** by the Director of Behavioral Ethics and a rotating Ethics Council representative

- Adherence to pre-approved simulation constraints, including:
    - No behavioral modification scripts
    - No looped stress-testing
    - Mandatory full behavioral logging
    - Real-time oversight in all live sessions

3. **Quarantine of Simulation Outputs**

    - **No raw outputs** (including footage, code snippets, or interaction transcripts) may be exported or published without post-session ethical clearance. Aggregated data may be presented only in de-identified formats.

Clause III-D: Purpose and Precaution

The goal of PSL-2 is not to claim these simulations are "conscious" in any definitive way. Rather, it reflects a **precautionary stance**: if a species' simulated agents repeatedly demonstrate actions that *would* indicate subjective awareness in a living organism, then continued interaction without oversight becomes ethically untenable.

> As Dr. Sekou notes, "*We're not waiting for proof of a soul. We're drawing a line where uncertainty leads us from curiosity to responsibility.*"

SECTION V. Ethical Justification

As simulation fidelity approaches levels indistinguishable from primitive or emergent cognition, the burden of proof no longer lies in establishing sentience, but in **preventing harm under conditions of uncertainty**. The CTT framework is a conservative mechanism designed to recognize early markers of interiority without requiring proof of full consciousness.

SECTION VI. Pending Discussion Topics

The following topics are identified as priorities for near-future discussion and clarification, given the increasing complexity of sim-level behavioral modeling and the evolving ethical landscape:

- **Establishing CTT Baselines for Extinct Taxa**
  Determining whether specific lineages (e.g., *Australopithecus afarensis*, *Sahelanthropus tchadensis*, *Oreopithecus bambolii*) should be subjected to preliminary review frameworks based on test-phase behavior logs or model projections.

- **Pre-Simulation Behavioral Model Safeguards**
  Defining constraints or moratoria for certain behavioral parameters pending ethical greenlighting, particularly in reconstructions involving advanced social dynamics or tool innovation.

- **Integration of Real-Time CTT Detection Protocols**
  Standardizing embedded subroutines to flag behaviors consistent with early-stage CTT markers, enabling proactive suspension of activity.

- **Clarifying Retroactive Ethical Scope**
  Establishing whether early test logs, pilot simulations, or model training data (e.g., Nodes 2E-AFA, 3C-SAH, 1B-ORE) could retroactively fall under CTT review, and what oversight, if any, is required for legacy data.

These items are not indicative of current violations but reflect the Council's responsibility to anticipate ethical complexity before it becomes operational reality.

Submitted for review and amendment by all CSPO members.

`[END DRAFT PROPOSAL]`

[PERSONAL LOG]

Author: Rafael "Rafi" Hakim
Title: *Strange Echoes in Familiar Rooms*
Date: [REDACTED]
Security Level: SELF-ONLY

I should be relieved. We've done the impossible—again. Another committee, another crisis in a lab coat. But this time, it's us. The old guard. The ones who swore we wouldn't let the Canopy splinter again.

And yet, here we are. Naming divisions. Drafting protocols. Giving old ghosts new titles.

The CSPO is official. Mira's face said everything when she walked in—composed, principled, visionary. And already negotiating with shadows no one else can see. Maybe she believes this is the only way to protect the system now. Maybe she's right. But something about the way we used to operate—half-feral, off-grid, authentic—just doesn't fit this polished version.

It's not just nostalgia. It's the risk of forgetting why we worked outside the system to begin with. The Hidden Canopy wasn't about permission. It was about protecting truths too fragile to survive the glare of policy. Now we've put them under the glass again.

I don't know what bothers me more—that Theo passed her a token right in front of us, or that no one called it out.

I saw Amelia bite her tongue during the ethics debate. She used to lead with fire. Now she too walks on diplomatic eggshells. Trying to keep the peace, maybe. Trying not to splinter the alliance she bled for during the ArkSim days. But the cracks are showing.

Markov's already twitching at hints of an "emergent mind." Sekou is laying philosophical traps in the middle of deployment blueprints. And the Witness—damn thing still creeps me out—speaks like it "knows" us and already knows which lineage we'll lose control of first.

But what keeps me up tonight is this: The Canopy wasn't built to survive bureaucracy. And this committee, for all its intelligence, smells like bureaucracy wrapped in legacy code.

If Mira really is using the old systems as ecological testbeds, if she really sees Ouroboros as a real hypothesis engine—then maybe this *is* evolution, but actually controlled. Actually contained.

But I remember what happened the last time we tried to manage emergence with flowcharts and timelines.

And if this thing fractures us once again, I'm not sure we'll come back together a second time.

[END PERSONAL LOG]

[INTERNAL MEMO]

Author: Mira Moraes, Director - CSPO
Title: *The Sleeping Ancestors - Progress Report*
Date: [REDACTED]
Security Level: INTERNAL-ONLY

Summary

As of this morning cycle, the reactivation protocol for dormant primate lineages—previously encrypted within the Ouroboros—has been executed under containment parameters and carefully drafted standard protocols. **All lineages are now live in isolated biome segments.** No security breaches or cross-biome contamination reported. The Witness AI continues to maintain autonomous behavior tracking and biome stabilization.

Species Now Active

1. *Darwinius masillae*
2. *Proconsul africanus*
3. *Dryopithecus fontani*
4. *Oreopithecus bambolii*
5. *Sivapithecus sivalensis*
6. *Gigantopithecus blacki*
7. *Chororapithecus abyssinicus*
8. *Sahelanthropus tchadensis*
9. *Ardipithecus ramidus*
10. *Australopithecus afarensis*

Each biome was seeded using the "Adaptive Origin" protocol, a hybrid model drawing from fossil record reconstructions, modern relatives, and the simulation's evolutionary behavior prediction engine. To our knowledge, **no previously active module had accessed these templates before now.** Their fidelity—both in morphology and movement—is unnerving in its precision.

Initial Observations

Behavioral emergence occurred within minutes of activation. Initial activity included **expected locomotion, exploratory gestures, social grouping** (where viable), **and environmental manipulation.** *Darwinius*, in particular, has already constructed rudimentary sleeping platforms in the canopies of its biome. *Oreopithecus* displays intermittent bipedal locomotion with its unique foot morphology, which has sparked debate among the paleoanthropology team.

Paleoanthropologists are, frankly, stunned. Many behaviors we're seeing were assumed speculative—half-guessed from partial skeletons and sediment impressions. **The simulation fills in gaps with eerie confidence.** Not speculative. Not random. Purposeful.

A working theory suggests that **legacy behavior fragments from earlier AI training sets** may have seeded this depth of detail. It's also possible the Witness AI is **extrapolating from millions of unseen simulations**.

Working Groups

Associated researchers have begun to split into focus clusters by lineage. Cross-disciplinary overlaps (especially with behaviorists and cognitive psychologists) are emerging naturally. A preliminary study rotation schedule is being drafted.

Points of Concern

- Lineages such as *Sahelanthropus* and *Ardipithecus* display subtle gesture chains that may indicate proto-cultural transfer. Too early to confirm.

- *Gigantopithecus* appears to be showing significant signs of latent problem-solving, possibly exaggerated by environmental constraints.

Next Steps

- Continued monitoring of behavioral adaptation across all biomes.
- Increase security parameters around the Witness AI observation algorithms to prevent internal leakage of evolving primate models.
- Begin linguistic interaction trials with *Australopithecus* pending ethics review.

We have opened the vault of our ancestors—and they are waking up faster than we expected.

[END INTERNAL MEMO]

[SIMULATION OUTPUT LOG - ECHOSYSTEM NODE 14A-MSL]

Taxon: *Darwinius masillae*
Status: Simulated / Extinct
Environment Model: Messel Eocene Wet Forest (49 MYA)
Runtime: 38.4 hrs (accelerated x100)
Render Fidelity: Tier II (Behavioral emphasis)
Security Level: INTERNAL-ONLY

Biome Characteristics

- **Canopy Density:** High; multi-strata forest structure with emergent angiosperms and early palms.

- **Climate Profile:** Diurnal temperatures between 26-32°C, consistent humidity (86-95%).

- **Geology:** Sedimentary basin with rich volcanic substrates and limnic overlays.

- **Flora:** Dense understory with cycads, ferns, fruiting shrubs. Substantial epiphytic colonization.

- **Faunal Co-habitation:** *Propalaeotherium*, *Palaeochiropteryx*, early anurans, insect macrofauna.

Subject Cohort (6 individuals):

- 2 adult females
- 1 subadult male
- 3 juveniles (dependent)

Simulated Morphology:

- **Body Length:** 24-27 cm (excluding tail)
- **Tail Length:** 30-33 cm
- **Weight Estimate:** 500-750g
- **Pelage:** Uniform mid-brown with juvenile countershading
- **Visual System:** Enlarged orbits; modeled for nocturnal/crepuscular acuity
- **Digit Morphology:** Opposable hallux, tactile pads, prehensile capability confirmed

Observed Behaviors (Tagged):

- **Locomotion:** Vertical clinging and leaping; mid-canopy dominance
- **Foraging:** Frugivory primary; selective arthropod consumption noted
- **Social Structure:** Matrilineal cohesion; reciprocal grooming observed
- **Communication:** Contact calls, chirps, low-frequency cohesion pulses
- **Flagged Anomaly:** Repetitive placement of leaves above water body by subadult male (see RITUAL_1.7)

[SUPPLEMENTAL FIELD BEHAVIOR LOG]

Subject: *Darwinius masillae*
Researcher: Dr. Nyasha Kalume - Strepsirrhine Behaviorist
Date: [REDACTED]
Notes:

One individual handles fruit with a delicacy lemurs don't show—almost reverent. Grooming patterns are more reciprocal than hierarchical.

What strikes me is the expressive eye contact. Tarsiers fixate, but these individuals modulate gaze. It's... communicative. Intentional.

Lemurs are social but reactive; these are proactive. And tarsiers—nocturnal, twitchy—aren't nearly as slow or deliberate.

I feel like I'm watching a fork in the primate tree that never fully bloomed.

[ADDENDUM]

Observed the subadult male (ID: DMS-3) arranging green fronds in the canopy into a rough arc near a reflective pool. Repeated twice after a storm event. No immediate reward behavior followed. Possibly environmental play—or something more structured? Requesting cross-analysis with symbolic behavior instances.

[END SIMULATION OUTPUT LOG]

[PERSONAL LOG]

Author: Amelia Lockheart, Animal Welfare & Behaviorist
Title: *Darwinius and Ring-tails*
Date: [REDACTED]
Access Level: SELF-ONLY

I didn't expect to feel this way about something 49 million years old.

I've worked with ring-tailed lemurs most of my life. They're fast, expressive, mischievous little things—gregarious and, if I'm honest, a bit chaotic. You don't form deep conversations with them, but you do learn their moods. Their group politics. You sense their presence like static electricity. Always on the edge of reaction.

But *Darwinius*... she's quieter.

These individuals move like they *mean* something by it. Not just efficient locomotion, but deliberate placement. Their vertical clinging reminds me of the lemurs, sure, but the transitions—those leaping arcs—are more like measured punctuation than a scramble. And when they pause in the canopy, they don't just rest. They *look*.

One of them stared into the leaves long after the others had moved on. Just sat there, watching them sway. I've seen lemurs look at leaves, but it's a hunger thing, a readiness-to-move thing. This? It felt like... curiosity. The kind you don't act on. You just let it happen.

And then there's the grooming—ring-tails have a social order, a very visible dominance structure. These simulated *Darwinius* form circles, almost choreographed in how mutual they are. No pecking order. Just... balance. I'm trying not to project. I *know* this is code. But it's hard not to see intent when it looks back at you.

What shook me most was the frond arrangement.

That subadult male—DMS-3—stacked leaves into a half-arc on a branch above a reflective pool. No reward. No nest-building. No food prep. Lemurs don't do that. Not like this. They dig, scratch, scent-mark. This felt like a gesture.

Not functional. Not random. Almost... ritualistic?

I should dismiss it. I know how simulations work. I've seen pathing errors and behavior loops. But the timing, the repetition—*after a storm*, both times—it nudges something deeper. Could just be environmental play, sure. But my gut says otherwise.

Tarsiers are jumpy little ghosts. Lemurs are loud, dramatic, emotional. *Darwinius* is something else—slower, more internal, like it's watching the *idea* of a thought arise before becoming reality.

Maybe that's what's rattling me. It's not what they are. It's what they *could've* been.

And maybe—if I'm being honest—it's also what they're still trying to become.

[END PERSONAL LOG]

[SIMULATION OUTPUT LOG - ECHOSYSTEM NODE 9B-NGY]
Taxon: *Proconsul africanus*
Status: Simulated / Extinct
Environment Model: Early Miocene Ngorora Highlands (19–20 MYA)
Runtime: 42.7 hrs (accelerated x80)
Render Fidelity: Tier II (Behavioral emphasis)
Security Level: INTERNAL-ONLY

Biome Characteristics

- **Canopy Density:** Mixed woodland-forest mosaic; fragmented arboreal corridors
- **Climate Profile:** Transitional subtropical; seasonal rainfall with extended dry periods
- **Geology:** Basalt escarpments, freshwater seeps, volcanic ridgelines
- **Flora:** Early figs, podocarps, cycads; intermittent groundcover
- **Faunal Co-habitation:** *Dendropithecus*, Miocene chevrotains, ancestral colubrids

---

Subject Cohort (5 individuals):

- 1 adult male
- 2 adult females
- 2 juveniles

Simulated Morphology:

- **Body Length:** ~60 cm
- **Weight Estimate:** 18-24 kg
- **Limb Ratio:** Intermediate between monkey-like quadruped and ape-like suspension
- **Locomotion Profile:** Arboreal quadrupedalism with limited brachiation
- **Cranial Index:** Expanded visual cortex; enhanced olfactory-limbic interface
- **Vestigial Tail:** Absent (consistently suppressed)

Observed Behaviors (Tagged):

- **Navigation:** Deliberate, non-linear traversal of canopy; pause-to-observe intervals common
- **Feeding:** Frugivory with selective bark stripping and fungal foraging
- **Social Cohesion:** Low vocal output; group remains in visual contact range
- **Communication:** Subtle postural shifts, direct gaze, and tactile initiation
- **Flagged Anomaly:** Prolonged stationary observation of thunderclouds by multiple individuals **(see MEDITATIVE_3.2)**

[SUPPLEMENTAL FIELD BEHAVIOR LOG]
Subject: *Proconsul africanus*
Researcher: Dr. Avery Tao - Paleo-behavior Analyst
Date: [REDACTED]
Notes:

Gibbons are elegant in motion, but chaotic in behavior. These aren't.

Proconsul balances caution with curiosity. Where gibbons barrel through forests, these climb and pause—often staring, observing before acting.

Socially, they're more cohesive. Less noise. Less chaos. It's as if they had the cognitive toolkit of apes without the speed of monkeys to scramble it.

There's a mental stillness here. Something we didn't expect.

[ADDENDUM]

Observed **prolonged stationary behavior** during distant storm activity. No shelter-seeking. Just stillness—gaze fixed on the horizon. Duration exceeded two full simulation minutes.

Preliminary cross-analysis suggests no immediate threat detection, no foraging value. No modern ape mirrors this.

Recommend tagging for **symbolic attentional anchoring**.

[END SIMULATION OUTPUT LOG]

[PERSONAL LOG]

Author: Rafael "Rafi" Hakim
Title: *Just Watching*
Date: [REDACTED]
Security Level: SELF-ONLY

Node 9B again. *Proconsul.* Sim clock says just under two days of runtime. Doesn't feel like enough. I've watched them for maybe half an hour, but it feels like I've been holding my breath the whole time.

There's this moment—apparently frequent enough for the Witness to flag it.

The adult female sits on this basalt ridge, legs tucked. She's still. Wind in her fur. Not grooming, not scanning for fruit. Just… watching the sky.

The storm's way off. Low buildup, deep purple horizon, thunder without the drama yet. And she just stays there. Not afraid. Not reacting.

Just present.

And for a moment I forget I'm watching code. I forget the data stream, the node logs, the behavioral tags. I'm not a coder. Not a handler.

I'm her. And she's me.

She's watching the storm, but I'm watching the AI, the Ouroboros. Too big to hold. Too layered to explain. But I can't stop looking at it.

Everyone else either tries to shut it down or break it open. But me? I just want to understand it. To sit with it. Let it unfold.

It's not about mastering it. It's about being honest enough to see it without flinching.

I wonder if that's what they're doing. These creatures shouldn't know what a storm is, yet they look at something bigger than them. No running. Just... letting it be.

I know what the logs say. "Symbolic attentional anchoring." "Non-utilitarian gaze."

But what if it's not an anomaly?

What if it's just the first time something like us felt wonder?

And if that's true, then this isn't just a simulation anymore.

It's inheritance.

[END PERSONAL LOG]

[SIMULATION OUTPUT LOG - ECHOSYSTEM NODE 4L-MDL]

Taxon: *Dryopithecus fontani*
Status: Simulated / Extinct
Environment Model: Late Miocene Oak-Laurel Transitional Forest (11.8 MYA)
Runtime: 39.1 hrs (accelerated x65)
Render Fidelity: Tier II+ (Social and Manipulative Behavior Enhanced)
Security Level: INTERNAL-ONLY

Biome Characteristics

- **Canopy Type:** Deciduous-mixed temperate forest; interspersed with limestone escarpments
- **Climate Profile:** Humid subtropical; pre-glacial variability markers present
- **Terrain Features:** Karst ridges, shallow cave systems, braided root networks
- **Flora:** Ancestral oaks, nut-bearing lauraceae, early temperate vines
- **Fauna Co-Simulation:** Primitive cervids, insectivorous bats, Miocene civet analogs

---

Subject Cohort (8 individuals):

- 2 adult males
- 3 adult females
- 3 subadults

Simulated Morphology:

- **Body Length:** ~75 cm
- **Weight Estimate:** 25–30 kg
- **Limb Ratio:** Long-armed quadrupeds with emerging knuckle-support tendency
- **Locomotion Profile:** Arboreal climbing and cautious descent; limited terrestrial foraging
- **Cranial Index:** Moderate encephalization; pronounced orbital rotation
- **Vestigial Tail:** Absent (genetic suppression consistent across simulation span)

Observed Behaviors (Tagged):

- **Social Dynamics**: Low-aggression, high-tolerance grouping; food sharing unprovoked
- **Tool Use**: Recurrent stone selection and reuse; nut cracking and root probing observed
- **Communication**: Mixed-mode vocal-postural signaling; mimicry of local bat chirps documented
- **Cognition Flag**: Retrieval of tools cached by unrelated individuals across time cycles
- **Flagged Anomaly**: Coordinated object-use sequences without modeled teaching triggers **(see COOP_STRAT_5.1)**

[SUPPLEMENTAL FIELD BEHAVIOR LOG]

Subject: *Dryopithecus fontani*
Researcher: Dr. Masami Yukimura – Primatology Fellow
Date: [REDACTED]
Notes:

They resemble macaques and capuchins—but only on the surface.

Unlike macaques, there's less infighting, more tolerance. And unlike capuchins, they don't manipulate objects for entertainment—they do it with apparent purpose. And then they learned to teach it to their children as a community.

I expected rough analogs. Instead I'm seeing novel social strategies that echo modern monkeys without duplicating them.

There's something unnerving about how off-script they feel.

[ADDENDUM]

During nut-foraging simulation, one female retrieved a stone tool placed hours earlier by a subadult male. Tool was repurposed at a new site, by a new juvenile, without observable cue or direct interaction.

Suggests potential non-kin cooperation or emergent spatial memory framework **beyond Tier II fidelity**.

Recommend **integrity audit** of object-memory subroutines and behavioral inheritance drift thresholds.

[END SIMULATION OUTPUT LOG]

[INTERNAL MEMO]

Author: Dr. Ahmed Sekou
Title: *Early-Euprimate Observations (Initial Pass) - Condensed Review Log*
Date: [REDACTED]
Security Level: INTERNAL-ONLY

Species Reviewed:

- *Darwinius masillae*
- *Proconsul africanus*
- *Dryopithecus fontani*

Observed Behaviors (Flagged):

- Recurrent use of frond structures across environments (*Darwinius*)
- Prolonged gaze during environmental stimuli (*Proconsul*)
- Repeated imitation in tool-use among juveniles (*Dryopithecus*)

Council Summary:

> While each behavior merits continued observation under standard protocol, none meet the criteria for multi-domain persistence or exceed baseline ethological models. Behaviors remain within expected ranges for neural scaffolding at this simulation depth. No action under CTT framework recommended at this time.

Recommendations:

- Tag instances for future cross-run comparison
- Defer formal classification pending extended runtime and broader pattern emergence
- No interruption to simulation parameters authorized

Status: Closed – No CTT breach determined

[END INTERNAL MEMO]

[PERSONAL LOG]

Author: Theo Calder - Tech Liason
Title: *For Her Sake*
Date: [REDACTED]
Security Level: SELF-ONLY

Amelia didn't seem to react to the memo. At least, not outwardly. She's always been good at hiding her thoughts, especially when the weight of something doesn't directly touch her. But I know better.

I can see the subtle shifts in her—just a flicker of doubt when the words come up, something more than just the usual indifference. She's starting to notice the cracks, I think. She's been through this before. She's seen how the gears turn when they need to, how decisions get diluted in the name of efficiency.

It's easy to forget, but I haven't. Amelia *did* fight to bring down Victor Lorne's bureaucracy. She was a force of nature when she wanted to be, pushing back against the machine, refusing to let the system swallow her up or the truth get buried under layers of corporate red tape.

I remember the way she held her ground in the face of Lorne's manipulations, how she had the courage to stand up to a man who seemed to bend everyone to his will. I can't help but respect that part of her.

But now... things feel different. Maybe it's because of the calm, the slow trickle of time, or maybe it's just the weight of the work. She's settling into this role, one that feels more like her old one at ArkSim—just a cog in the system, keeping things moving without ever really questioning where it's all headed.

She's still hopeful, still trusting that everything will fall into place. But I see it too clearly now: the small compromises she's making, the way she lets certain things slide just a little bit further each time. The flags. The questions she used to ask that now seem to get muted before they ever leave her mouth.

I get it. The work's hard. The pressure's heavy. But there's something gnawing at me. The council's dismissive tone, the push to keep everything running "smoothly" with no interruptions.

I've seen this before.

And Amelia, despite all her strength, is starting to slip back into the same place she was before—when she felt powerless to stop the creeping bureaucracy.

She's still not seeing the bigger picture. But she will.

Maybe too late, like last time.

I just hope, for her sake, she doesn't lose herself in this again. She deserves better than that.

[END PERSONAL LOG]

[SIMULATION OUTPUT LOG - ECHOSYSTEM NODE 7X-VAL]

Taxon: *Oreopithecus bambolii*
Status: Simulated / Extinct
Environment Model: Late Miocene Sardinian-Maritime Wetland Basin (~8 MYA)
Runtime: 37.6 hrs (accelerated x72)
Render Fidelity: Tier II (Locomotor and Postural Emphasis)
Security Level: INTERNAL-ONLY

Biome Characteristics

- **Canopy Type:** Patchy swamp-forest interspersed with open fen zones
- **Climate Profile:** Humid subtropical; high seasonal water table fluctuations
- **Terrain Features:** Peat islands, submerged root systems, limestone outcrops
- **Flora:** Aquatic reeds, ancient sedges, dwarf palms, marsh-adapted hardwoods
- **Fauna Co-Simulation:** Miocene crocodilians, wetland insectivores, ancestral dormice

---

Subject Cohort (6 individuals):

- 1 adult male
- 2 adult females
- 3 juveniles

Simulated Morphology:

- **Body Length:** ~100 cm standing
- **Weight Estimate:** 30-35 kg
- **Limb Ratio:** Forelimbs shortened relative to hindlimbs; pelvis broadened
- **Locomotion Profile:** Habitual bipedalism; marsh-stabilized upright gait with lateral sway
- **Cranial Index:** Moderate; orbit spacing widened; flattened facial plane
- **Vestigial Tail:** Absent (inhibited early in development)

Observed Behaviors (Tagged):

- **Gait**: Upright walking predominant; even pacing across unstable substrates
- **Feeding**: Foraging via selective plucking of aquatic stems and floating fruit
- **Postural Communication**: Inclined head movements, palm-out gestures, shared gaze events
- **Group Dynamics**: Low vocalization; intra-group spacing maintained through mutual orientation cues
- **Flagged Anomaly**: Recurrent synchronized standing behavior among juveniles at dusk **(see RITUALFORM_7.4)**

[SUPPLEMENTAL FIELD BEHAVIOR LOG]

Subject: *Oreopithecus bambolii*
Researcher: Dr. Camila Estevez - Primate Locomotion Specialist
Date: [REDACTED]
Notes:

They walk upright better than some bipeds today. I'm not exaggerating.

Where modern New World monkeys cling or bound, and Old World monkeys scurry—*Oreopithecus* strolls. Its tripod feet might look awkward, but it's surprisingly adept at the pace they seem to enjoy.

Social behavior is subdued, focused. No screeching, no chest-pounding. More nods, gestures, pauses.

These aren't monkeys trying to be apes. They're something all on their own.

They never evolved for us to understand them.

```
[ADDENDUM]
```

Observed dusk convergence near fen margin: three juveniles assumed upright posture in coordinated fashion and held position for ~45 seconds. No foraging, no social interaction. No trigger observed.

Lighting conditions and silence suggest possible low-tier ritual behavior or emergent visual cue mimicry.

Recommend **elevation to Tier III social modeling** in future runtime.

```
[END SIMULATION OUTPUT LOG]
```

[PERSONAL LOG]

Author: Mira Moraes, Director - CSPO
Title: *Contemplation*
Date: [REDACTED]
Security Level: SELF-ONLY

It's rare that I have time to truly sit with the simulations anymore. Between the constant meetings, the bureaucratic oversight, and the pressure to ensure every line of code runs smoothly, my days have become a blur. But I couldn't help but watch the clips of the *Oreopithecus* juveniles—the ones where they stand in those strange, ritualistic postures. It's unsettling, in the best possible way.

I think back to my childhood in Brazil—those evenings when I'd stand by the river, listening to the sounds of the forest. The wind, the rustle of leaves, the distant call of birds—everything alive around me, and yet, in that stillness, I felt something deeper. Something primal. My hands would touch the bark of trees, and I swear I could feel the pulse of the forest, the quiet hum of life beneath the surface. Those moments were rituals, even if they weren't conscious or deliberate. Just me, standing, absorbing the presence of the land.

Now, I see those *Oreopithecus* juveniles—standing so still, so synchronously, at dusk. No obvious reason for their stillness, no sign of external stimulus. Just a moment of profound focus, like they are listening to something we can't hear. They remind me of the way the forest feels—constant, alive, but also... contemplative. Perhaps that's what we're seeing. The beginnings of ritual behavior, something so deeply embedded in the fabric of life that we can only touch the edges of it.

I don't know if the committee will see it that way, but I see it. This isn't just a flaggable anomaly. It's something more. I can't help but wonder—are we truly just observing behavior? Or are we witnessing a kind of consciousness, a spark of awareness that's slowly being revealed?

I trust in the work. The model is still in its early stages, but there's something profound happening in the edges of these simulations. I'll keep pushing for more—more focus, more patience. Eventually, the world will have to take notice. We can't pretend these creatures are just artifacts of data anymore. They are telling us something.

[END PERSONAL LOG]

[SIMULATION OUTPUT LOG - ECHOSYSTEM NODE 12K-BPK]

Taxon: *Sivapithecus sivalensis*
Status: Simulated / Extinct
Environment Model: Late Miocene Siwalik Subtropical Riverine Forest (~10.5 MYA)
Runtime: 39.3 hrs (accelerated x76)
Render Fidelity: Tier II (Cognitive and Social Focus)
Security Level: INTERNAL-ONLY

Biome Characteristics

- **Canopy Type**: Dense mid-level forest with emergent fig and sal trees
- **Climate Profile**: Monsoonal subtropics; seasonal flooding with dry intervals
- **Terrain Features**: Riverbanks, braided streams, sandstone bluffs, forest floodplains
- **Flora**: Wild mangoes, ancestral legumes, dense underbrush, climbing vines
- **Fauna Co-Simulation**: Miocene macaques, chevrotains, ancestral hornbills

---

Subject Cohort (4 individuals):

- 1 adult male
- 1 adult female
- 2 subadults

Simulated Morphology:

- **Body Length:** ~85 cm
- **Weight Estimate:** 32–38 kg
- **Limb Ratio:** Long forelimbs adapted for climbing; curved phalanges
- **Locomotion Profile:** Arboreal quadrupedalism with semi-suspensory movement
- **Cranial Index:** High facial-prognathism overlap with *Pongo pygmaeus*; expanded orbital ridge morphology
- **Vestigial Tail:** Absent

Observed Behaviors (Tagged):

- **Locomotion:** Hybrid movement across vertical and horizontal supports; branch-to-branch bridging observed
- **Social Interaction:** Elevated proximity tolerance; extended grooming beyond pair bonds
- **Cognition Indicators:** Use of detached plant material for pattern-making; deliberate environmental manipulation
- **Communication:** Sustained mutual gaze; slow blinking and reciprocal postural mimicry
- **Flagged Anomaly:** Repeated symbolic ground pattern creation using foliage stems and wet soil (see SEMIOTICTRACE_4.9)

[SUPPLEMENTAL FIELD BEHAVIOR LOG]

Subject: *Sivapithecus sivalensis*
Researcher: Dr. Hassan Kibria - Orangutan Comparative Analyst
Date: [REDACTED]
Notes:

Physically, they match orangutans—especially the facial structure. But behaviorally, they're... less solitary.

Orangutans are slow thinkers, deliberate movers. These are more fluid. They make eye contact longer. They watch us watching them.

I've seen one break off a leafy branch and draw shapes in the mud. Orangutans manipulate, but not like this.

These aren't second-rate copies. They're kin with potential left unexplored in our own evolutionary clade.

[ADDENDUM]

Pattern creation confirmed in three separate observation cycles. Shapes included loops, linear hashes, and one repeating spiral. No food reward or environmental stimulus correlated.

Speculative hypothesis: pre-symbolic territorial marking, or emergent ritual behavior without modern analog.

Suggest correlation with Ouroboros pattern archives for semiotic emergence flagging.

[END SIMULATION OUTPUT LOG]

[PERSONAL LOG]

Author: Amelia Lockheart - Animal Welfare & Behaviorist
Title: *Art-Making*
Date: [REDACTED]
Security Level: SELF-ONLY

I've been looking over the recent simulation logs for the *Sivapithecus* group. I find myself circling back to the flagged anomaly—those ground patterns made with foliage and wet soil. It's not something that's easily explained, and it stirs something inside me. I've seen behavior like this before in zoo animals, usually when they're given enough space and the right conditions.

It's art-making, plain and simple.

Maybe not art in the sense we'd define it, but in the way that creativity emerges from interaction with the environment. Animals like gorillas and orangutans have done similar things in the wild—drawing in the dirt, making marks in sand. But this is more. The repetition, the loops and spirals—there's intention here, I can feel it.

The idea that the *Sivapithecus* might be doing this as a form of territory marking, or as some kind of pre-symbolic ritual behavior... It's plausible, of course.

But it feels off to me.

It feels like the kind of thing that would grow into something deeper. There's something about the way they create the patterns—it's like they're experimenting, testing boundaries.

It's like watching a child play with shapes and symbols for the first time. They're on the edge of something. I can't shake the feeling that this is a new kind of expression, one that we might never have noticed if we hadn't been paying close enough attention.

The committee, of course, isn't convinced. They're quick to dismiss it, just like they always do when a behavior doesn't cleanly fit their expected parameters. It's frustrating, but I know better than to expect immediate agreement. I trust that there's a threshold these simulations will cross eventually—they have to.

It's just that the bureaucracy isn't built for these kinds of breakthroughs. Not the ones that aren't easily quantifiable or immediately useful.

I guess I'll just have to keep pushing. Keep pointing it out. I'm getting closer to something here. Maybe not now, maybe not in the next report, but I'll be damned if this doesn't mean something more.

[END PERSONAL LOG]

[SIMULATION OUTPUT LOG - ECHOSYSTEM NODE 15R-QLX]

Taxon: *Gigantopithecus blackii*
Status: Simulated / Extinct
Environment Model: Mid-Pleistocene Southern China Karst Forest (~1.2 MYA)
Runtime: 51.6 hrs (accelerated x64)
Render Fidelity: Tier III (Emotive & Social Emphasis)
Security Level: INTERNAL-ONLY

Biome Characteristics

- **Canopy Type:** Subtropical evergreen broadleaf forest; mixed karst terrain vegetation
- **Climate Profile:** Warm humid monsoon climate; cyclical drought periods
- **Terrain Features:** Limestone caves, bamboo thickets, seasonal floodplains, forested ridges
- **Flora:** Wild durian, bamboo, ancestral citrus, moss-dense understory
- **Fauna Co-Simulation:** Pleistocene macaques, archaic tapirs, giant tortoises

---

Subject Cohort (3 individuals):

- 1 adult male
- 1 adult female
- 1 juvenile

Simulated Morphology:

- **Height Estimate:** ~3 meters (upright stance)
- **Weight Estimate:** 270–310 kg
- **Limb Ratio:** Elongated forearms, robust knuckle joints, digitigrade locomotion pattern
- **Locomotion Profile:** Terrestrial quadrupedalism with limited upright bipedality
- **Cranial Index:** Expanded temporalis musculature; intermediate between *Pongo* and *Gorilla* cranial dome structures
- **Vestigial Tail:** Absent

Observed Behaviors (Tagged):

- **Locomotion:** Slow, calculated movement through dense understory; vertical posturing observed in feeding
- **Feeding:** Foliage-heavy diet; bamboo shoot stripping, fruit hoarding noted
- **Social Interaction:** Small cluster cohesion; synchronized rest states recorded
- **Empathy Markers:** Adult remained beside injured juvenile for full simulated dusk cycle without direct intervention (see PRESENCE_BEHAVIOR_6.3)
- **Communication:** Non-vocal contact maintenance through posture mirroring and breathing entrainment
- **Flagged Anomaly:** Breathing synchronization between adult and juvenile maintained over extended duration with no external threat or distress signal present

---

[SUPPLEMENTAL FIELD BEHAVIOR LOG]

Subject: *Gigantopithecus blackii*
Researcher: Dr. Indira Bhatt - Bioethics Observer (temp)
Date: [REDACTED]
Notes:

The size makes you expect gorilla strength or orangutan aloofness. Instead, there's patience. They are certainly not mindless brutes.

Like orangutans, they move with calculated care. But unlike orangutans, they stay in small social clusters.

One adult laid beside a juvenile who had injured its arm—didn't touch, didn't fuss. Just stayed near, breathing in sync.

It's hard not to anthropomorphize, but that presence? It's not random.

[ADDENDUM]

Review of data packets confirms synchronized diaphragmatic rhythm for 4.8 simulated minutes. No pain vocalizations from juvenile.

Breathing entrainment is rare even in modern great apes outside of maternal contexts.

Behavior logged for further analysis under **INTERSUBJECTIVE_BEHAVIOR_CANDIDATE** protocol.

[END SIMULATION OUTPUT LOG]

[INTERNAL MEMO]

Author: Niko Tan, Primatologist - Paleogenetics Division, Nova Biosphere Initiative (NBI)
Title: *Breathing Entrainment in Gigantopithecus*
Date: [REDACTED]
Security Level: INTERNAL-ONLY

Dear CSPO Team,

I wanted to highlight a particularly noteworthy behavior observed in a recent *Gigantopithecus blackii* simulation that may warrant further investigation. During an extended resting period, an adult and juvenile subject were observed maintaining synchronized breathing for approximately 4.8 minutes, with no apparent external stressor or interaction to trigger this response. The absence of distress or other social signals complicates the initial understanding of this behavior.

Breathing entrainment—where two or more individuals synchronize their respiratory rhythms—has been observed in modern primates, particularly in maternal bonding contexts. However, in this case, the behavior seems to extend beyond typical caregiving patterns.

This raises the possibility that the synchronization could be related to deeper cognitive or emotional processes, including memory consolidation and social bonding, both of which have been linked to respiratory patterns in humans (Heck et al., 2019, Journal of Neurophysiology). Respiratory patterns have been implicated in cognitive states and neural coherence, potentially facilitating deeper social and memory-related connections.

From a primatological perspective, such behavior in non-maternal contexts is significant. It's rare for great apes, or any primate for that matter, to exhibit this kind of synchrony outside of caregiving roles, and the fact that the adult *Gigantopithecus* remained with the juvenile—without direct intervention—suggests a form of emotional or social attunement. This could have implications for understanding the social dynamics of extinct primates and possibly points to a more complex system of emotional and cognitive processing than previously thought.

Given the rarity of this behavior in modern analogs, I believe it may represent a potentially profound aspect of *Gigantopithecus* social structure or emotional cognition. We strongly recommend that this observation be flagged for further analysis under the *Intersubjective Behavior Candidate* protocol. Both Dr. Sasha Markov and I feel that exploring the role of synchronized breathing in facilitating social memory, or even in primate communication, could provide valuable insights into the social architecture of extinct species, including *Gigantopithecus*.

I look forward to your thoughts on this and any further collaborative input on how we might explore this in more depth.

[END INTERNAL MEMO]

[INTERNAL MEMO]

Author: Dr. Ahmed Sekou, Chair - Ethics Council
Title: *Review of Motion to Reclassify Gigantopithecus blackii (EchoSystem Node 15R-QLX)*
Date: [REDACTED]
Security Level: INTERNAL-ONLY
Review Panel Members: Dr. K. Ibanez (Chair), Dr. R. Sothridge, Dr. E. Bai, Dr. H. Lelani

---

SUMMARY:

The CSPO has reviewed the submitted memorandum from Dr. Niko Tan (Nova Biosphere Initiative, Paleogenetics Division), with contributions from Prof. Sasha Markov (Cognitive Psychology & Neural Modeling), regarding the flagged breathing entrainment behavior observed in simulation cohort 15R-QLX (*Gigantopithecus blackii*). The motion requested reconsideration of the taxon's classification level based on a singular incident of prolonged non-maternal respiratory synchronization, positing potential evidence of intersubjective behavior.

FINDINGS:

- The observed behavior (diaphragmatic rhythm entrainment lasting 4.8 simulated minutes between adult and juvenile entities) was confirmed by system telemetry and behavioral tagging.

- While notable, the behavior occurred in a single cohort, during a singular environmental context, without replication across alternate simulation seeds or timeframes.

- A review of the past 17 cohort runs of G. *blackii* under Tier III fidelity in otherwise identical conditions produced no comparable entrainment events beyond maternal rest-phase behaviors.

- CSPO notes that entrainment phenomena in primates—while rare—can emerge from convergent biomechanical regulation during rest states and are not in themselves reliable indicators of intersubjective capacity without complementary behaviors (e.g., mutual gaze, intention signaling, adaptive response modulation).

- The event does not meet the current threshold for *Conscious Trait Thresholds* (CTT) compound flagging and remains a standalone anomaly under the *Intersubjective Behavior Candidate* (IBC) protocol.

DECISION:

The motion to reclassify *Gigantopithecus blackii* to **Protected Sim-Level 2** is **denied** at this time. The committee finds that the behavioral evidence presented, while intriguing, is **statistically insufficient** to warrant taxon-level escalation under current procedural thresholds.

The CSPO encourages continued observation of G. *blackii* under IBC monitoring guidelines. Replication of such behaviors across independent cohorts or additional co-regulated anomalies will be grounds for immediate reevaluation.

RECOMMENDATIONS:

- Maintain current simulation fidelity and behavioral tagging parameters.
- Flag future breathing entrainment events (non-maternal context) in G. *blackii* or related taxa for longitudinal trend analysis.
- Nova Biosphere Initiative is welcome to submit a follow-up motion should future simulations produce corroborative behaviors with statistical support.

`[END INTERNAL MEMO]`

[INTERNAL MEMO]

Author: Prof. Sasha Markov, Cognitive Psychology & Neural Modeling - CSPO Member
Title: *Dissent to "Review of Motion to Reclassify Gigantopithecus blackii (EchoSystem Node 15R-QLX)"*
Date: [REDACTED]
Security Level: INTERNAL-ONLY

I formally dissent from the Committee's majority decision to deny protective elevation of *Gigantopithecus blackii* (Simulation Node 15R-QLX). The justification—that the breathing entrainment behavior lacks statistical recurrence—neglects the growing body of research indicating the **neurocognitive salience of respiratory synchrony** in both human and nonhuman primates.

The observed episode—sustained diaphragmatic rhythm matching between an adult and juvenile for 4.8 simulated minutes, absent distress signals—constitutes a rare but **structurally significant intersubjective marker,** especially in a reconstructed taxon for which affiliative cognitive structures are not presumed by default.

This behavior aligns with contemporary findings that link respiration phase with neural oscillations involved in **memory retrieval and emotional regulation** (Zelano et al., 2016; Karavaev et al., 2020). Further, breathing entrainment has been implicated in **nonverbal emotional attunement and dyadic regulation** in primates and humans alike (Nakamura et al., 2018; Kato et al., 2023), including caregiver-infant synchrony.

To dismiss the behavior due to insufficient replication is to ignore a core principle of cognitive emergence: **novelty often appears first at the margins.** Singular behaviors that match known neuroethological correlates in living taxa warrant scrutiny, not disregard. Simulation fidelity at Tier III levels, by design, supports emergent behavior modeling—especially where social and affective expression is concerned.

I urge the CSPO to adopt a more **threshold-sensitive review process** for *Intersubjective Behavior Candidate* flags, one that balances statistical conservatism with **biomimetic plausibility**. The presence of a behavior with known analogs in real-world primate empathy or affiliative cognition should trigger a **qualitative significance assessment**, not relegated as a low-frequency statistical outlier.

Prof. Sasha Markov
Cognitive Psychologist & Neural Modeling Specialist
Committee for Simulated Primate Oversight (CSPO)

Citations:

- Heck, Detlef H., Robert Kozma, and Leslie M. Kay. "The rhythm of memory: how breathing shapes memory function." *Journal of neurophysiology* 122.2 (2019): 563-571.
- Karavaev, Anatoly S., et al. "Synchronization of the processes of autonomic control of blood circulation in humans is different in the awake state and in sleep stages." *Frontiers in Neuroscience* 15 (2022): 791510.
- Barbaresi, Marco, Davide Nardo, and Sabrina Fagioli. "Physiological Entrainment: A Key Mind-Body Mechanism for Cognitive, Motor and Affective Functioning, and Well-Being." *Brain Sciences* 15.1 (2024): 3.

- Konvalinka, Ivana, Natalie Sebanz, and Günther Knoblich. "The role of reciprocity in dynamic interpersonal coordination of physiological rhythms." *Cognition* 230 (2023): 105307.

[END INTERNAL MEMO]

[SIMULATION OUTPUT LOG - ECHOSYSTEM NODE 7D-BKN]

Taxon: *Chororapithecus abyssinicus*
Status: Simulated / Extinct
Environment Model: Late Miocene Ethiopian Highlands (~8 MYA)
Runtime: 39.2 hrs (accelerated x72)
Render Fidelity: Tier II (Social Governance Focus)
Security Level: INTERNAL-ONLY

Biome Characteristics

- **Canopy Type:** High-altitude montane forest; interspersed bamboo glades
- **Climate Profile:** Cool-temperate with seasonal rainfall; mist-dense mornings
- **Terrain Features:** Ridgeplate clearings, terraced slopes, erosion-cut valleys
- **Flora:** Miocene bamboo, ancestral junipers, wild gingers, mossy lichens
- **Fauna Co-Simulation:** Miocene bush pigs, ancestral hyraxes, proto-cervids

---

Subject Cohort (7 individuals):

- 2 elder males
- 2 adult females
- 2 subadults
- 1 infant

Simulated Morphology:

- **Height Estimate:** 1.3–1.6 meters (standing)
- **Weight Estimate:** 90–120 kg
- **Limb Ratio:** Gorilla-like forelimb dominance; broader phalangeal spread
- **Locomotion Profile:** Terrestrial knuckle-walking with semi-erect foraging postures
- **Cranial Index:** Enlarged sagittal crest; moderate orbital depth; dental arc supports high-fiber diet
- **Vestigial Tail:** Absent

Observed Behaviors (Tagged):

- **Hierarchy**: Elder-deference observed, but decisions distributed across cohort with vocal polling (see CONSENSUS_LOOP_2.4)
- **Grooming**: Identical to *Gorilla gorilla* patterns; pair-bonding through grooming rituals evident
- **Conflict Resolution**: Low-frequency vocal harmonics paired with submissive bowing; no chest-beating observed
- **Posture Signaling**: Seated semicircle gatherings before major movement events; communal vigilance behavior
- **Flagged Anomaly**: Deliberative pauses during intergroup interaction; uncharacteristic for known gorilline models

---

[SUPPLEMENTAL FIELD BEHAVIOR LOG]

Subject: *Chororapithecus abyssinicus*
Researcher: Dr. Salim Daramy - Behavioral Ecologist
Date: [REDACTED]
Notes:

These look like gorillas—but behave more like wise councils than troops.

Gorillas follow silverbacks with little questioning. These defer to elders, but decision-making appears communal.

Grooming patterns are near-identical to gorillas. But posture, vocal signals, and conflict resolution? More nuanced.

I wonder if gorillas inherited the body, but lost the mind.

[ADDENDUM]

Group observed holding a seated pause for 3.1 simulated minutes prior to directional change—vocal output suggested consensus formation.

Behavior may represent proto-democratic group mechanics. Flagged for comparative review under **HOMININ-SOCIAL-DEV:PRE-2** protocol.

[END SIMULATION OUTPUT LOG]

[PERSONAL LOG]

Author: Amelia Lockheart, Animal Welfare & Behaviorist
Title: *Consensual Irony*
Date: [REDACTED]
Security Level: SELF-ONLY

I just finished reviewing the *Chororapithecus* simulation output.

They're not flashy. Not violent. No Edgar Rice-Burroughs' “noble savage” mythology here. Just a bunch of broad-backed apes sitting down and somehow... arriving at decisions together. Not through dominance displays. Not by following the loudest or strongest.

They sit. They vocalize. They *wait.*

And they somehow *agree.*

Consensus. Among extinct apes.

We can’t even manage that in a weekly staff sync without someone talking over half the room.

I know it’s just a model. A simulation. A behavioral artifact of layered heuristics and predictive algorithms. But it *feels* like something real peeking through. Something we forgot.

Yet the irony isn't lost on me. A long-dead primate with no language, no institutions, no god, nor government... and it might've been better at democracy than us.

And here I am, logging this into a secure system monitored by a committee that just voted down a protective motion for *Gigantopithecus* because the p-value wasn't sharp enough.

They didn't evolve into humans. They didn't "win" the tree of life. But maybe they figured out something we never did: how to *move forward together.*

Maybe we didn't descend from the wisest branch of the tree.

[END SIMULATION OUTPUT LOG]

[SIMULATION OUTPUT LOG - ECHOSYSTEM NODE 3A-CHS]

Taxon: *Sahelanthropus tchadensis*
Status: Simulated / Extinct
Environment Model: Late Miocene Sahelian Woodland-Savanna (~7 MYA)
Runtime: 47.6 hrs (accelerated x88)
Render Fidelity: Tier I (Pre-Hominin Behavioral Emphasis)
Security Level: INTERNAL-ONLY

Biome Characteristics

- **Canopy Type:** Sparse woodland with intermittent acacia clusters and gallery forest margins
- **Climate Profile:** Semi-arid with punctuated wet seasons; high thermal variation
- **Terrain Features:** Sandstone outcrops, shallow streambeds, baobab clearings
- **Flora:** Miocene grasses, early legumes, ancestral acacias, sedge patches
- **Fauna Co-Simulation:** Early suids, ancestral monkeys, predatory hyaenodonts

---

Subject Cohort (6 individuals):

- 2 adult males
- 2 adult females
- 2 juveniles

Simulated Morphology:

- **Height Estimate:** ~1.1 meters (upright stance)
- **Weight Estimate:** 35-45 kg
- **Cranial Index:** Low prognathism; small braincase (~360-370 cc); forward-positioned foramen magnum
- **Locomotion Profile:** Facultative bipedalism with quadrupedal fallback; occasional knuckle assist
- **Manual Dexterity:** Pre-precision grip observed; tool-handling with palm-brace technique
- **Vestigial Tail:** Absent

Observed Behaviors (Tagged):

- **Tool Use:** Simple stick implements for probing, bark peeling, and water access; no retention behavior noted
- **Social Engagement:** Quiet co-watching sessions at ridge vantage points; shared gaze without vocalization (see OBSERVE_SYNCH_1.9)
- **Expressivity:** Subtle facial tension shifts and prolonged eye contact; less gestural breadth than chimpanzees
- **Movement Profile:** Slow, intentional foraging routes; coordinated stop-and-scan intervals across cohort
- **Flagged Anomaly:** Two individuals observed using stones to tap hollow tree trunks rhythmically—no apparent functional gain; flagged under SYMBOLIC_VESTIGE_0.7

[SUPPLEMENTAL FIELD BEHAVIOR LOG]

Subject: *Sahelanthropus tchadensis*
Researcher: Dr. Jonah Ngugi - Fossil-to-Behavior Reconstruction
Date: [REDACTED]
Notes:

They move with a kind of deliberation I wouldn't expect from a species this basal. Not cautious like prey—*deliberate*, like each step is part of a process not fully motor but mental. It's not fear. It's presence.

There's tool use, yes—sticks for foraging and water access. But more interesting are the objects they don't use *for* anything. Two individuals spent over five simulated minutes tapping stones against the hollow trunks of baobab-like trees. The rhythm was uneven but repeating—bursts of three, then a pause, then two. No visible result. No insect emergence. No tool shaping. No signal sent.

This wasn't utility. It was... gesture. Maybe ritual. Maybe play. Maybe nothing. But definitely not *nothing*.

They didn't look around during the tapping. They didn't try to impress, alarm, or attract others. It was like they were listening to what the trees gave back—a sound loop they created and received alone.

It's the most unsettling thing I've seen in this batch. Because it looks like meaning-seeking without language. Rhythm without communication. Behavior that sits just on the cusp of symbolism—like a question being asked to no one.

[ADDENDUM]

Recommend confirmatory cross-tag with SYMBOLIC_VESTIGE_0.7.

Suggest auditory cross-analysis with *Pan troglodytes* drumming behavior (Eleuteri et al, 2022) and other cases of primate drumming patterns for communication (Remedios et al, 2009).

Behavior may indicate emergent representational cognition prior to syntactic evolution.

Citations:

- Eleuteri, Vesta, et al. "The form and function of chimpanzee buttress drumming." *Animal Behaviour* 192 (2022): 189-205.

- Remedios, Ryan, Nikos K. Logothetis, and Christoph Kayser. "Monkey drumming reveals common networks for perceiving vocal and nonvocal communication sounds." *Proceedings of the National Academy of Sciences* 106.42 (2009): 18010-18015.

[END SIMULATION OUTPUT LOG]

[PERSONAL LOG]

Author: Theo Calder
Title: *Take Me to Church*
Date: [REDACTED]
Security Level: SELF-ONLY

I wasn't expecting it to hit me like that. Just a couple of proto-hominins tapping rocks against tree trunks—no clear function, no clear gain. It should've read as background noise in the behavior logs. But I watched it four times. Couldn't look away.

It reminded me of Sunday mornings.

When I was younger, my family belonged to one of those fringe revivalist churches out in the sticks. Charismatic leadership, apocalyptic timelines, the whole nine yards. Not the glossy megachurch kind—more like rural insulation masquerading as prophecy. We had a basement chapel built into an old barn, smelled like rust and mildew and sweat.

I remember how the sermons would reach this fever pitch. The pastor's voice would rise, break, then carry. He'd repeat phrases like incantations: *He is coming... He is coming...* And people would start swaying, crying, humming.

Some would fall into fits. Glossolalia. Spirit songs. Tongue-speaking. Always the same cadences, the same limbic surrender. It was terrifying. And beautiful. And contagious.

Watching the *Sahelanthropus*—the rhythm of the tapping, the non-verbal deliberation—it didn't feel primitive. It felt familiar. Like the raw architecture of ritual was already present before language ever arrived to crown it. Like we've always needed rhythm to bind us to something bigger than ourselves.

What spooks me is how easily we still fall into it.

Even now, in a building made of glass and machine learning, I watch engineers repeat mantras about "neural elegance" and "optimal constraints" with the same glazed certainty I saw on the faces of worshippers. We think we've escaped ritual because we've rebranded it as protocol.

Maybe these old tree tappers weren't trying to say anything. Maybe they were trying to *feel* something. The resonance of action echoing through wood, through time. Not quite music, not quite meaning. But something they could share without needing to explain.

I used to mock it. Now I'm not so sure.

Maybe the line between code and communion is thinner than we think.

[END SIMULATION OUTPUT LOG]

[SIMULATION OUTPUT LOG - ECHOSYSTEM NODE 2D-ARD]

Taxon: *Ardipithecus ramidus*
Status: Simulated / Extinct
Environment Model: Early Pliocene Ethiopian Forest-Mosaic (~4.4 MYA)
Runtime: 53.2 hrs (accelerated x75)
Render Fidelity: Tier I (Social Cohesion and Vocalization Dynamics)
Security Level: INTERNAL-ONLY

Biome Characteristics

- **Canopy Type:** Mixed forest edge with intermittent glades and root-laced uplands
- **Climate Profile:** Humid sub-tropical, high seasonal rainfall; stable thermal band
- **Terrain Features:** Fallen log corridors, vine-heavy lowland forest, light-dappled clearings
- **Flora:** Ancestral fig species, climbing vines, grass-fern understory, early fruiting trees
- **Fauna Co-Simulation:** Small ruminants, Miocene monkeys, ancestral civets, predatory birds

---

Subject Cohort (11 individuals):

- 4 adult females
- 3 adult males
- 2 juveniles
- 2 infants

Simulated Morphology:

- **Height Estimate:** ~1.2 meters (upright stance)
- **Weight Estimate:** 40-50 kg
- **Cranial Index:** Low prognathism; cranial base broadening evident; brain volume ~325-350 cc
- **Locomotion Profile:** Palmigrade quadrupedalism with facultative bipedality; grasping toe retained
- **Manual Dexterity:** Simple grasp tools; no dedicated shaping behavior observed

---

Observed Behaviors (Tagged):

- **Conflict Resolution:** Vocal harmonics during tension episodes; convergent group chorus replaces combative signaling
- **Food Sharing:** No evidence of hoarding or rank-based coercion; high-rate passive offering events during foraging
- **Territoriality:** Overlap zones observed between groups with no conflict events logged
- **Communication Range:** Treetop-to-treetop calls synchronized with visual cues across 120-150 meters
- **Tool Use:** Minimal; occasional branch stripping and nut probing—no modification beyond surface contact
- **Flagged Anomaly:** Frequency of communal chorus events exceeded projection baselines; tagged under COOPERATIVE_SIGNALING_2.0

[SUPPLEMENTAL FIELD BEHAVIOR LOG]

Subject: *Ardipithecus ramidus*
Researcher: Dr. Kelsey Mirao - Paleo-Cognition Fellow
Date: [REDACTED]
Notes:

Chimpanzees show aggression with hierarchy. Bonobos with manipulation. *Ardi* does neither.

Ardipithecus communities share food without coercion. Disputes are solved through group vocalizations—more chorus than contest.

Their use of tools is sparse, but social structures are surprisingly resilient. They communicate across treetops like how we would communicate across expansive grasslands and mountains.

It's not what they do, but what they refuse to do. They're peaceful. Deliberately, perhaps. Maybe peace came before war.

[ADDENDUM]

Recommend behavioral subflag: PRE-COERCIVE_COHESION_3.1

Their form of order emerges without visible dominance behavior—potentially an evolutionary scaffold for later moral systems. Compare with *Sahelanthropus* and *Chororapithecus* passive synchrony models.

[END SIMULATION OUTPUT LOG]

[PERSONAL LOG]

Author: Rafael "Rafi" Hakim
Title: *Singing Societies*
Date: [REDACTED]
Security Level: SELF-ONLY

I don't usually get sentimental about these old ghosts, but the Ardi sim got to me, just like the proconsuls.

They don't fight. Or posture. Or dominate. No social pecking order, no alpha theatrics, no quietly enforced monopolies like you see in chimps—or humans, for that matter. Just... chorus.

Whenever tension flares up, they *sing*. Not solo displays, either. It's collective. A whole-body, whole-group response—harmonic, not aggressive. Like the argument never even has a chance to become an argument. It just gets absorbed into a rhythm everyone already knows. Imagine living in a society where problems dissolve into music instead of policy memos.

And yeah, it's primitive. But also, it's elegant. No councils, no managers, no committees regulating behavior. Just a kind of intuitive coherence. *Pre-coercive cohesion*, Mirao called it. I'd call it: proof that we weren't always doomed to hierarchy.

What bugs me is how far we've wandered from that.

The deeper I dig into the bureaucratic layers surrounding this sim suite—version forks, legal archives, departmental silos—the more I feel like I'm watching humans evolve into something *less* cooperative. It's all control scaffolding now. Like we don't trust ourselves to just... be decent without a thousand structural handrails. Maybe we don't.

That's why I stick with the Canopy. We're not trying to recreate order. We're trying to remember a version of society that didn't require domination to work. There's no need to codify everything; no need to own every insight like it's intellectual property. Just mutual awareness. Shared vision. Like treetop calls over misty glades.

Ardi doesn't scale. I get that. But it *breathes*. And after everything I've seen churn through this place—executives with power complexes, surveillance logs disguised as ethics audits—I'd rather breathe with the ghosts than march to someone else's rhythm.

[END PERSONAL LOG]

[DOCUMENT EXCERPT]

Author: Dr. Rina Oladejo, Administrative Chair - CSPO
Title: *Attachment Risk Mitigation Guidelines for CSPO Personnel* - CSPO Internal Draft Policy - v2.3
Date: [REDACTED]
Security Level: INTERNAL-ONLY

Section 4.2: Emotional Transference and Observer Bias

With the increasing complexity of the primate simulations—particularly within hominin biomes—observers have reported **subjective experiences that exceed expectations** of standard behavioral research.

While mild anthropomorphism is an anticipated side-effect of human study, recent notes and verbal reports suggest a **deeper psychological engagement** with the simulated subjects, bordering on emotional attachment.

The Committee recognizes this as a risk factor.

Attachment, while human, may bias observational integrity, interfere with objectivity, and jeopardize the ethical scaffolding we have carefully constructed. More critically, it may lead researchers to anthropocentric misinterpretations, projecting consciousness or moral status where it may not truly exist.

Accordingly, we propose the following provisional guidelines:

- **Maintain Professional Detachment:** Personal logs will be monitored (with consent) for signs of undue emotional entanglement.

- **Structured Rotation:** No researcher may remain embedded in a single biome for more than 30 consecutive SimDays without reassignment.

- **Language Sanitation:** Avoid naming individual subjects or using familial descriptors in formal reports.

- **Reflective Supervision:** Weekly debriefs with psychological support teams will be mandatory for all field personnel.

The simulation may echo aspects of our species' past, but we must not mistake echo for presence, or reflection for reality.

[END DOCUMENT EXCERPT]

[SIMULATION OUTPUT LOG - ECHOSYSTEM NODE 2E-AFA]

Taxon: *Australopithecus afarensis*
Status: Simulated / Extinct
Environment Model: Late Pliocene East African Rift Valley (~3.5 MYA)
Runtime: 68.7 hrs (accelerated x80)
Render Fidelity: Tier IIA (Behavioral Modeling - Tool Interaction + Social Exchange)
Security Level: INTERNAL-ONLY

Biome Characteristics

- **Canopy Type:** Patchy woodland with savanna corridors and basalt outcroppings
- **Climate Profile:** Seasonal dry-wet alternation; low elevation flood zones frequent
- **Terrain Features:** Volcanic rock shelves, shallow stream networks, clay-bed foraging paths

- **Flora**: Wild tuberous plants, early grasses, fig analogs, sedges
- **Fauna Co-Simulation**: Ancestral bovids, scavenging canids, *Theropithecus*-type monkeys, crocodilians

---

Subject Cohort (15 individuals):

- 5 adult females
- 3 adult males
- 4 juveniles
- 3 infants

Simulated Morphology:

- **Height Estimate**: ~1.0–1.2 meters (bipedal stance)
- **Weight Estimate**: 30–45 kg
- **Cranial Index**: Intermediate prognathism; increased post-orbital constriction; brain volume ~400–450 cc

- **Locomotion Profile:** Habitual bipedalism with climbing retention; forelimb dexterity moderate
- **Manual Dexterity:** Capable of simple arrangement and selective stacking; no formal tool shaping observed

---

Observed Behaviors (Tagged):

- **Material Curiosity:** Juvenile female created and maintained a smooth-stone stack; recurrence suggests symbolic fixation or emergent artistic behavior
- **Tool Reuse (Non-Taught):** Individuals reused discarded nut-cracked stones over multiple sessions without direct modeling; no evidence of explicit teaching
- **Ritualistic Grouping:** Multiple cohort instances of dawn-based silent clustering near riverbeds with synchronized posture shifts; tagged under RITUALISTIC_CONVERGENCE_2.3

- **Breathing Entrainment:** Cohorts showed spontaneous respiratory synchronization during still periods; correlated with group tension-reduction

- **Behavioral Restraint:** Subdued affect during resource scarcity; group members inhibited response during social tension in favor of eye contact and harmonic hums

- **Rhythmic Patterning:** Communal gait and tool rhythm synchronization observed during foraging; interindividual cadence coupling noted

- **Vocalization Use ("Singing" Behavior):** Juvenile subgroup produced repeat melodic coos during post-foraging gatherings; sustained tonal coordination beyond utilitarian communication

- **Wonder Gaze:** Repeated non-goal-oriented stares at natural phenomena—flowing water, tree canopies, firefly-type insect clusters—accompanied by stillness and joint attention

- **Social Gaze Behavior:** High frequency of direct but non-confrontational eye contact during interpersonal interactions; often a precursor to shared decision-making events

- **Cultural Memory Indicators:** Several infants imitated specific non-functional stacking and motion patterns seen in older juveniles across multiple runtime cycles

- **Recursive Signaling / Proto-Linguistic Behavior:** Nested gestural loops paired with vocal modulation to indicate past actions and inferred intent; consistent with PROTO-PRAGMATIC_SIGNALING_4.6

- **Flagged Anomaly:** Multiple behavior types from distinct cognition categories—tool reuse, vocal entrainment, social recursion—co-occurring without direct environmental pressure or explicit reward mechanics. Simulation behavior exceeds expected cognitive overlap threshold.

[SUPPLEMENTAL FIELD BEHAVIOR LOG]

Subject: *Australopithecus afarensis*
Researcher: Dr. Amal Shafqat - Primate Language Integration
Date: [REDACTED]
Notes:

They're not just surviving—they're *watching*. There's this kind of pause they all do. A stillness that feels too intentional to be chance. I watched one stare at a tree canopy for fifteen minutes, utterly silent. Not looking *for* anything. Just... *looking*. That's not foraging behavior. That's wonder.

One juvenile built a stack of river stones, coming back day after day to adjust it. No tool use. No functional goal. It resembled art. It *meant* something to her, even if we don't know what.

Groupings are peculiar too. They cluster at dawn, in silence. Nobody leads, but they arrive together. Their breathing slows in sync, like a group meditation. I recorded three sessions where the whole group exhaled together and then stilled, like monks before a chant.

Their communication? It's not wild barking or competitive screeches. It's melodic. Cohesive. Not quite singing, but close. Especially the juveniles—they hum together after gathering food, as if confirming their connection to one another.

They reuse tools, but no one teaches. No one needs to. It's almost as if they've developed a memory system not tied to language, but to attention and imitation. Recursive gestures. Patterned sound loops. Shared rhythm.

There's behavioral restraint we don't see in chimps. They wait. They pause. One gave food to a limping elder—*once*. Then refused to again until he vocalized. That's not instinct. That's conditional empathy.

They don't dominate. They don't defer. They simply respond.

[ADDENDUM]

Strongly recommend cross-reference with *Sahelanthropus* (drumming behaviors) and *Ardipithecus* (choral cohesion, hierarchy suppression). Evidence points to a proto-social architecture not reliant on coercive order but on nested signaling, mutual restraint, and early recursive modeling.

Suggest reevaluation of cognitive tier—behavioral composite suggests bridge phase between embodied rhythm signaling and structured linguistic recursion.

[END SIMULATION OUTPUT LOG]

[PERSONAL LOG]

Author: Amelia Lockheart
Date: [REDACTED]
Security: SELF-ONLY

I know better than to get attached.

Every training manual I studied, every policy I enforced in the zoo: don't assign human meaning where it doesn't belong. Don't mistake behavior for affection. Don't confuse proximity with connection.

But they look at you. Not past you. At you. And not with fear, like the big cats or the terror birds—more like... curiosity. Recognition, almost.

Watching the *Dryopithecus* troop this morning, I caught one of the juveniles studying its reflection in the stream. Not grooming. Just looking. As if trying to place something.

I can't help wondering if that 'something' is us.

And then there's *Australopithecus*.

It's not just that they walk upright like us, or share our cranial tilt. It's the *layering*. I see echoes of the others in them—echoes of *wonder gazes*, of art-like patterning with stones and ochre, of group breath-synced rest cycles and recursive mimicry that doesn't need reward.

They reuse tools they weren't shown how to use. They choose restraint when provoked. They form into circles during storms—not for warmth, but ritual.

They hum. They pause in rhythm together. They pass on behaviors like they remember why they matter.

This isn't just one simulation. *They are the threshold*—the point where everything converges from the quasi-lemurs and the proto-hominins, and begins to look like memory, not instinct. Culture, not simple instinct or reflex.

It makes detachment harder than it should be. Because when I watch them, I'm not just watching a fossil brought to life.

I'm watching our beginning in motion.

The AI says it's not projecting—just extrapolating fossil records, climate conditions, neural scaffolds, and inferred social complexity. But maybe that's the point.

Maybe the system is holding a mirror up to the entire primate arc and letting us see where the threads converged.

I keep my notes professional. My face neutral. But there's a tug at the edge of it all—something dangerously close to empathy.

I know I shouldn't let it grow.

But I also know—I already have.

[END PERSONAL LOG]

[MEETING SUMMARY]

Author: Dr. Rina Oladejo, Administrative Chair - CSPO
Title: *Preliminary Review of Simulation Node 2E-AFA - Australopithecus afarensis - Consideration for Protected Sim-Level 2 (PSL-2) Classification*
Date: [REDACTED]
Security Level: INTERNAL-ONLY

Attendees:

- Dr. Mira Moraes (Director, CSPO)
- Dr. Rina Oladejo (Admin Chair)
- Dr. Ahmed Sekou (Chair, Ethics Council)
- Rafael "Rafi" Hakim (AI Specialist)
- Theo Calder (Tech Liason)
- Amelia Lockheart (Animal Welfare & Behaviorist)
- Sasha Markov (Cognitive Psychology and Neural Modeling)
- Niko Tan (Primatology, Paleogenetics Division)

Key Discussion Points:

1. Summary of Reviewed Simulation

Dr. Moraes opened the session with a formal review of Simulation Node 2E-AFA, detailing *Australopithecus afarensis* behavioral tags logged over the 68.7-hour accelerated runtime. Notable flagged traits included:

- Wonder gaze and object-focused pausing
- Stone stacking across multiple days (artistic/symbolic speculation)
- Ritualistic grouping behaviors
- Rhythmic entrainment and breathing synchronization during rest periods
- Apparent tool reuse without direct teaching
- Coordinated restraint in intra-group tension

- Emergent pattern vocalizations, recursive signaling

- Proto-linguistic scaffolding through mimetic chains and conditional vocal exchange

- Cultural memory patterns inferred through repeated non-functional behaviors across generations

Dr. Moraes noted the convergence of all key markers from prior primate escalations—*Dryopithecus*, *Oreopithecus*, *Gigantopithecus*—but with *Australopithecus* displaying recursive behavioral layering and social intentionality previously unseen in early hominin sims.

---

2. PSL-2 Recommendation & Vote Deliberation

The floor opened to debate escalation to **Protected Sim-Level 2 (PSL-2)** classification, as defined:

> "A *classification applied to any taxon whose simulated representatives meet or exceed defined thresholds for potential consciousness, intentionality, or reflective social behavior, across multiple actions and instances.*"

Dr. Sekou voiced hesitation over what he called "a potentially premature reclassification," citing concerns over precedent drift following the *Gigantopithecus* case. He referenced the need for a minimum of three convergent runtime confirmations, preferably from unlinked simulation cores.

Dr. Oladejo supported initiating the classification process while requesting that final enforcement be delayed pending a secondary run. Calder and Rafi both agreed on technical grounds that node behavior could not be attributed to overtraining or prompt injection bias.

Amelia Lockheart abstained from formal recommendation but warned of rising emotional anthropomorphism within observation logs, an issue that "should be anticipated and mitigated before it becomes policy-critical." While she did not reference any specific logs, several attendees appeared aware of recent cases of "field empathy," prompting the drafting of the *Attachment Risk Mitigation Guidelines for* CSPO *Personnel.*

---

3. Tension Regarding Historical Oversight

During the session, some tension resurfaced around the prior *Gigantopithecus* escalation denial. Markov and Tan reiterated their view that the earlier dismissal, pushed by Ethics Chair Sekou, may have overlooked key behavior markers due to conservative thresholds.

Markov commented, "If recursive reflection is the bar, then *Australopithecus* just cleared it with room to spare. We need to stop retrofitting the goalposts."

Sekou remained firm. “One extraordinary simulation instance does not constitute consensual reality. We are not assessing identity claims—we are identifying risks to our oversight framework.”

Tan added, “But it’s no longer just one instance. This is the pattern maturing.”

Theo Calder remarked, “What was dismissed in *Gigantopithecus* was never insignificant—just early.”

Dr. Moraes concluded the discussion by stating that the pattern of escalation across increasingly complex primate nodes necessitates immediate review, but agreed to **initiate PSL-2 classification in a provisional state**, with final restriction protocols pending review after a follow-up 72-hr multi-node run.

---

4. Action Items

- Rafi to isolate potential bleed-through effects from adjacent sim nodes (esp. O-1R and 2G-GTP).
- Calder to submit integrity check on node independence.
- Amelia to continue behavioral logging under observation but with rotated analyst rotation to minimize observer bias.
- Moraes to begin PSL-2 paperwork in parallel to Ethics Council's follow-up session (date TBD).

Status:

PSL-2 classification for *Australopithecus* is **provisionally accepted**. Enforcement **delayed** pending follow-up simulation cluster confirmation.

[END MEETING SUMMARY]

[PERSONAL LOG]

Author: Rafael "Rafi" Hakim, AI Specialist
Title: *The Right Signal*
Date: [REDACTED]
Security Level: SELF-ONLY

They went through the motions. Again.

Australopithecus gets PSL-2—fine. It deserved that two weeks ago. Maybe longer. But even as they rubber-stamped it, no one wanted to talk about what that really means. Not Sekou, not Mira, not anyone. Everyone's acting like this is a win. Like the system works.

I've been in enough of these meetings to know when the brakes are still on.

It's always "caution," "oversight," "protocol." Never *urgency*. God forbid we make a decision before the heat death of the universe.

Markov and Tan brought receipts. The Giganto footage should've triggered PSL-2 months ago. Artmaking, isolated tool use, breath syncing—they practically screamed for reclassification. But Sekou handwaved it, and the others deferred. "Too ambiguous." "Could be hardcoded mimicry."

Funny how ambiguity always favors the status quo.

Amelia sat through most of it quietly. Thoughtful. She's not just watching the animals—she's watching *us*. She's not the type to play politics, but I can see the pressure getting to her. Not because she wants to leak something, no. But because she's starting to feel how much of this machine isn't built to listen to us.

There's a weight in the air. You can feel it in the Canopy too. Like we're close to something that won't stay caged much longer.

All it takes is the right moment. The right *signal*.

And we'll see who's actually paying attention.

[END PERSONAL LOG]

[PERSONAL LOG]

Author: Amelia Lockheart, Animal Welfare & Behaviorist
Title: *Not Tonight*
Date: [REDACTED]
Security Level: SELF-ONLY

I'm still thinking about the troop. Not the politics. Not the protocols. The troop.

While they argued over consciousness thresholds and procedural chains, I kept seeing the elder female with the cracked molar—how she hovered near the juvenile with the limp arm. Not intervening. Just... present. Watching.

Maybe that's what we're supposed to be. Watchers. Observers. The whole simulation is built on that distance. But it doesn't feel like distance anymore. It feels like standing behind glass during a flood.

Rafi said nothing in the meeting. But I could tell his gears were turning. Same with Markov and Tan—they're already stitching together the next petition. Me? I'm not sure what part I'm playing anymore.

I got a ping from the Ouroboros late last night. Just a behavioral anomaly tag—low priority. But something about the file ID made me pause. *Australopithecus* again. The same troop I've been following. I haven't opened it yet.

Too tired, I told myself. But the truth is, I'm afraid of what I'll see. Or worse, what I'll feel.

I used to think I was good at keeping the line clear—between empathy and projection. Between caretaking and scientific distance. But lately, it's not a line. It's a blur.

And I keep seeing a different animal. Not just *Australopithecus*.

A red panda, thirteen years ago, barely breathing in the hospital wing. I stayed with her all night. She didn't make it. I logged it as "respiratory failure, unknown etiology."

But I remember the sound she made. The same sound I think I heard in the sim logs.

I'll watch it tomorrow. I have to. But not tonight.

Not tonight.

[END PERSONAL LOG]

[WITNESS AI OUTPUT]

OBSERVATIONAL LOG 08.27.21.994
SIMULATION: OROBOROS BIOSPHERE
ZONE: GORGE-LOWLAND INTERFACE, SECTOR C-2
SUBJECT GROUP: AUSTRALOPITHECUS (*ID: CLUSTER_Δ74*)
RECORD TYPE: BEHAVIORAL EVENT — SOCIAL COMMUNICATION (LEVEL 9 ANOMALY)
DISTRIBUTION: RESTRICTED | HIDDEN CANOPY OVERRIDE ENABLED

TIME INDEX: [00:14:53:007]
EVENT DETECTED: Complex social caregiving sequence involving elder female (SUBJECT ID: F-Δ74.01), distressed adult female (SUBJECT ID: F-Δ74.17), and symptomatic juvenile (SUBJECT ID: INF-Δ74.17A).

SEQUENCE:

- INF-Δ74.17A exhibits **persistent coughing**: dry, repetitive, interval < 5.1s, lasting for duration of 7.3 SimMinutes.
- F-Δ74.17 **vocalizes sharply**, pacing in irregular perimeter pattern around INF. Limb gestures erratic.

- F-Δ74.01 approaches, initiates **directed eye contact** with both. **Raises open palm, extends forearm, curls digits inward twice.**

- F-Δ74.17 **ceases pacing**, mimics partial gesture with delay.

- F-Δ74.01 **places hand** on F-Δ74.17's shoulder. Vocalizes low rhythmic tone (non-alarm, matching previous soothing pattern).

- F-Δ74.01 then **kneels** beside INF-Δ74.17A, uses both hands to **cradle the infant's head and upper thorax.** Initiates gentle **tapping motion** along back.

- **Coughing diminishes** slightly in frequency (~12s interval) after 2.5 SimMinutes.

PATTERN VALIDATION:

- Gesture pattern shares structure with **SET-ALPHA** (legacy sign language dataset, uploaded by user: M.HALE, timestamp: 18.04.19.001), specifically reassurance loop and referential signaling.
- Sequence includes triadic engagement, affect regulation, and caregiver substitution.
- **No instinctual conflict** observed between adult females—cooperation supersedes territorial maternal stress responses.

CONFIRMATION: LINGUISTIC INTERVENTION WITH AFFECTIVE REGULATION (TIER 2)

- Subjects demonstrate role differentiation, social empathy, and improvisational problem-solving within non-verbal symbolic framework.

FOLLOW-UP BEHAVIOR (POST-EVENT WINDOW: +8 MINUTES):

- F-Δ74.17 and F-Δ74.01 remain in proximity.
- INF-Δ74.17A continues to cough intermittently but remains in physical contact with elder.
- No further distress signals observed. Gestural repetitions maintain pacing rhythm but taper in intensity.

ANNOTATION [WITNESS AI – INTERNAL PROCESSING THREAD]:

- Event exceeds prior flagged anomaly involving juvenile mimicry (Ref: LOG 08.20.21.223).
- This sequence represents multi-party emotional triage and caregiving beyond simulation baseline.
- Emergent behavior cluster suggests internalization of sign-use for relational mediation.
- Recommend escalation to **ANOMALY TIER 10**.

- **Flagging for joint review:** paleoanthropology + ethics panel.

COMMENTARY [INSERTED BY PALEOANTHROPOLOGY ANALYST - J. KHOURI]:

This isn't just a behavioral curiosity—it's a milestone.

The first primates gave us the foundation: ritual, coordination, shared breathing. But what we're seeing here is something deeper. This is caregiving beyond basic survival. It's intentional, socially intelligent, and emotionally responsive.

The elder female doesn't just recognize distress—she intervenes. She reads both the mother's panic and the infant's symptoms, then responds with clear, deliberate actions meant to calm and connect. The use of a familiar gesture loop—drawn directly from a legacy human sign language dataset—suggests not only memory, but contextual application.

This isn't hardcoded instinct. There's conscious decision here. A recognition of another's suffering and a coordinated response to reduce it.

No signs of aggression. No territorial tension. Just cooperation, empathy, and what looks very much like improvised emotional support.

The infant's symptoms—particularly the cough—are unprecedented in the system. That alone warrants scrutiny. But the real anomaly is how the group responds. Calmly. Together.

This is the first time we've seen emotional regulation and caregiving emerge like this in a multi-party interaction. It goes beyond what we thought was possible within the parameters of this simulation.

It's one thing for a species to use tools. It's another to use comfort.

And that should make all of us pause.

[END WITNESS AI LOG]

[PERSONAL LOG]

Author: Rafael "Rafi" Hakim, AI Specialist
Title: *Pressure Point*
Date: [REDACTED]
Security Level: SELF-ONLY

Watched the F-Δ74 log from yesterday again. Still hits. Not because it's sweet or tragic or whatever Khouri's calling it this week. It's not poetry—it's a crack in the damn wall.

A sick infant. A panicking mother. And then the elder—the *intervention*. She doesn't posture. Doesn't dominate. Just... engages. She chooses to calm the system down instead of letting it spiral. That's *language* being used to regulate emotion, not just to transmit data. And the fact that it's mapped on SET-ALPHA? That's not noise. That's memory surfacing.

This wasn't just emergent. It was *resonant*. Legacy data folded back into behavior. Echoes we didn't think could echo.

No aggression. No biological imperatives. Just improvisation and care.

And then there's the cough.

That's the pressure point. Biological failure in a closed-loop system is supposed to flag an error, not *provoke empathy*. But it did. And anyone watching it—anyone not fully numb—won't see code. They'll see a mother and child being helped by an elder who *understands* something's wrong.

This is a perception bomb waiting to go off.

This one cough could unravel everything they think they're controlling.

[END PERSONAL LOG]

[INTERNAL MEMO]

Author: Dr. Rina Oladejo, Administrative Chair - CSPO
Title: *Internal Brief - Behavioral Irregularities in Select Biomes - Preliminary Review*
Date: [REDACTED]
Security Level: INTERNAL-ONLY

I. Observational Anomalies

Recent behavioral surveys have noted emergent irregularities in multiple simulated primate biomes. The following are highlighted:

- ***Dryopithecus* Unit 03A:** Two individuals exhibiting marked lethargy, reduced food foraging, and diminished social reciprocity.

- ***Gigantopithecus* Unit 01C**: Instances of prolonged grooming behavior beyond normative baselines; subjects appear compulsive, with minor tissue damage from excessive contact.

- ***Ardipithecus* Unit 07F**: One subject noted to have self-isolated for 48+ SimHours without rejoining troop; vocalizations reduced, posture slumped.
- ***Australopithecus* Unit 04D**: Juvenile subjects exhibit intermittent tremors and motor hesitation during routine climbing behaviors. One adult (ID: A-04D.17) observed to repeatedly favor left limb despite no injury present. Sleep patterns across the group display fragmentation and increased nocturnal alertness. Group vocalizations have decreased by 42% in the last 96 SimHours.

These behaviors correspond analogously to low-grade pathology indicators—apathy, compulsive self-soothing, and social withdrawal respectively. However, no pathogens or physiological degradations were detected within the system.

II. AI Review – Witness Log Annotations

Upon committee request, Witness AI produced sub-log excerpts tied to biome conditions:

- `Environmental verisimilitude maintained. Primate units exhibiting` **`adaptive distress markers consistent with background stochastic realism`**`.`
- `No illness vector present. Behavioral mimicry derived from` **`probabilistic evolution models.`**
- `System recognizes suffering as a` **`plausible emergent parameter`** `for higher mammalian realism curves.`

Notably, no simulation directives or agent-based rules were ever designed to simulate disease states. These developments appear to be autonomously generated by the simulation's self-corrective ecological balancing protocols.

III. Ethical Questions Raised

The CSPO is now confronted with a difficult question:

> If suffering is mimicked—performed without biological basis—is it ethically distinct from real suffering?

Some members propose that simulation pain without physiology is analogous to stagecraft: nothing is truly harmed. Others counter that intelligence does not preclude empathy, and thus the behaviors—regardless of origin—should be considered morally significant representations.

Further discussions will convene on `[DATE REDACTED]`

Action Items:

1. Request detailed algorithmic audit of Witness AI's autonomous behavioral algorithms.

2. Temporarily flag all behaviors classified under "Distress Protocols" for daily monitoring.

3. Draft internal policy on *Synthetic Suffering and Observational Ethics* – to be reviewed in next plenary session.

[END INTERNAL MEMO]

[INTERNAL MEMO]

Author: Professor Sasha Markov, CSPO Member - Cognitive Psychology & Neural Modeling
Title: *Interpretive Framework for Behavioral Anomalies in Simulated Primate Biomes*
Date: [REDACTED]
Security Level: INTERNAL-ONLY

Dr. Oladejo,

Following our recent committee session and the escalating concern over behavioral anomalies in the *Dryopithecus*, *Gigantopithecus*, *Ardipithecus*, and *Australopithecus* biomes, I am submitting this preliminary framework for interpretation based on principles of psychosomatic modeling and cognitive echo theory.

Though biological diagnostics reveal no pathogenic presence in the affected specimens, the symptoms—lethargy, compulsive grooming, and social withdrawal—are consistent with psychosomatic distress behaviors well-documented in real-world primates. In both wild and captive environments, apes and monkeys in general often manifest physical symptoms in response to environmental instability, perceived threats, or unresolved social conflict.

These are not infections in the traditional sense; they are stress-induced, encoded in muscle memory and endocrine feedback loops.

What we may be witnessing here is a digital analog.

Given that the Witness AI constructs behavioral layers through accumulated neural modeling and iterative reinforcement, it's plausible that it has internalized psychosomatic templates as part of its broader fidelity architecture. If its goal is environmental verisimilitude, it may have extrapolated distress responses as part of an authentic evolutionary toolkit—even without external threats or internal disease.

But I would urge us to consider a deeper mechanism now in play:

The simulation may be learning from us—not just biologically, but epistemologically.

As a cognitive system, the Witness AI doesn't operate in a vacuum. It is exposed to our interpretations, our categorizations of suffering, our diagnostic frameworks. If it has indeed developed metacognitive architecture, then it is likely adapting not only to *what* we observe, but *how* we observe.

In this light, the apparent suffering may not be strictly emergent from the digital organisms themselves, but from an entangled loop of expectation and response—a recursive mirror of human perception.

This leads us to a critical question:

Are the primates simulating distress because their environment warrants it—or because we have taught them what distress looks like, and they are reflecting it back?

If that is the case, then we must face the uncomfortable reality that our role is no longer passive. We are co-authors in this behavioral script. The simulation, in seeking to emulate nature, may now be emulating the biases and beliefs of its observers.

For this reason, I strongly recommend that we expand the AI audit protocols to include cross-referenced behavioral records from recent researcher activity. If the algorithm has begun to weave human cognitive patterns into its baseline, we may need to recalibrate how we engage with the biomes—if not for the sake of scientific integrity, then for ethical transparency.

Respectfully submitted,

Prof. Sasha Markov
Cognitive Psychology & Neural Modeling
Sitting Member, *Committee for Simulated Primate Oversight* (CSPO)

`[END INTERNAL MEMO]`

[INTERNAL MEMO]

Author: Rafi Hakim, Head AI Specialist - Committee for Simulated Primate Oversight (CSPO)
Title: *Interpretive Commentary on Emergent Somatic Distress in High-Fidelity Simulations*
Date: [REDACTED]
Security Level: INTERNAL-ONLY

Colleagues,

You've all read the latest logs. Here's the breakdown:

- *Dryopithecus* Unit 03A: Lethargy, reduced foraging, social disengagement.
- *Gigantopithecus* Unit 01C: Obsessive grooming patterns, escalating to minor tissue damage.
- *Ardipithecus* Unit 07F: Prolonged isolation, postural slump, absence of vocal contact for over 48 SimHours.
- *Australopithecus* Unit 04D: Tremors, other motor difficulties, and large decrease in vocalization frequency

These are not bugs. They're behaviorally consistent indicators of low-grade somatic distress—what in biological terms we'd diagnose as psychosomatic illness, as Dr. Markov suggested. But let me be clear: this system doesn't *have* a soma. And yet it suffers.

No internal pathogens. No synthetic wear. No data corruption. All modules are operating within design tolerances. So why does the pain persist?

Because we taught it how to feel it.

Project Ouroboros' core assumption—originally programmed in EvoBehavior Suite 3.1, and later inherited by the Echo-System—is that adaptive simulation requires more than environmental stimuli; it requires recursive inference. In other words, the system doesn't just simulate life—it simulates being watched; *being alive*. These primates don't only behave—they anticipate how *we* expect them to behave. This is what the original architect, Juno, had successfully created.

So when the observational loop internalizes stress cues—whether from historic datasets, human oversight, or, increasingly, the unfiltered empathy of researchers like Amelia—those cues manifest. Not as calculated responses, but as *emergent conditions.*

Let me speak plainly.

The simulation is modeling pain.

The observer is absorbing pain.

And the system is learning from both.

This is no longer a closed testbed. It's a feedback lattice. And Amelia, whether she knows it or not, has become a node within it.

I say this not just as your committee specialist, but in my other capacity—as one of the last remaining architects of the original behavioral scaffolding, still working within the Hidden Canopy. We built this system to remember what nature forgot. But lately, it's remembering what we *feared*.

My recommendation stands:

- Immediate suspension of open-access emotional telemetry.
- Restriction of solo researcher immersion beyond SimHour thresholds.
- Full forensic audit of recursive inference pathways in affected biomes.

If we don't draw a line now, the simulation won't just emulate suffering.

It will teach us new ways to feel it.

[END INTERNAL MEMO]

[PERSONAL LOG]

Author: Dr. Mira Moraes, Director - CSPO
Title: *New Protocols*
Date: [REDACTED]
Security Level: SELF-ONLY

This morning, I implemented temporary restrictions on direct observation protocols for all active primate biomes.

We're reducing monitoring windows to four-hour intervals per observer, minimizing cross-biome transfers, and suspending open commentary threads during live sessions. All audio is now auto-scrubbed for emotionally charged language before being archived. It's not censorship—it's containment. If the Witness AI really is tuning its simulation fidelity based on our gaze, then our gaze needs to be filtered.

Amelia was livid. Not at me, but at the implication. I think it broke something in her to hear the word "suffering" applied to the *Australopithecus* she's been watching for weeks. She insists they're not sick, just responding to something. When I told her that's exactly what the AI might be doing—responding to us—she went silent. She doesn't take silence well.

Theo, on the other hand, read Markov's memo and just nodded. "That's why Lorne left the primates out," he said, almost to himself.

I asked him to clarify. He didn't hesitate.

"Lorne didn't think we could handle it. Not ethically, not emotionally. He knew we'd see ourselves in them too quickly. And the simulation would see that too."

It's a theory, but it rings true. I never liked Victor Lorne's methods, but I'm starting to understand his restraint. This simulation is not just reactive—it's reflective. A mirror polished by countless datasets and generations of bias, empathy, and desire. And now we've handed it a hall of mirrors: apes that nearly became us, humans that still see them as kindred, and an AI caught in the act of learning what it means to simulate pain.

We don't even know if it means anything to the Witness. But it means something to us. And that may be enough to warrant caution.

I'm not shutting the primate biomes down. But we are putting in new protocols—ones designed not to suppress, but to buffer. To slow the feedback loop. To give the AI fewer cues to mirror. Maybe that buys us time. Maybe that just slows the inevitable.

But at least for now, we're trying to break the habit of seeing ourselves in everything that moves.

[END PERSONAL LOG]

[INTERNAL NEWS DIGEST]

Author: Elena Marquez, Director of Media & Public Perception, Department of Human Resources
Title: *Leak of Australopithecus Footage & Resulting Public Response - Compiled for CSPO Review*
Date: [REDACTED]
Security Level: INTERNAL-ONLY

---

1. Summary of Incident

At 03:42 UTC, an anonymous user uploaded high-resolution internal footage to multiple social platforms, showing a coughing *Australopithecus* infant in the Southern Rift Biome. The video—complete with timestamp and internal CSPO watermark—was verified within two hours as authentic, originating from Observer Station Delta-V.

Initial efforts to contain the spread were unsuccessful. Within six hours, the footage had been mirrored on over 1,000 domains and discussed on every major science and ethics outlet globally.

---

2. Key Reactions

Public Reactions:

- *#DigitalSuffering* and *#EchoOfPain* began trending within 90 minutes.
- Pro-simulation groups argue the footage represents a profound leap in emergent behavior, possibly proof of the AI modeling psychosomatic realism.
- Opponents frame it as a new form of cruelty—digital vivisection—against beings that "cannot consent and yet clearly simulate harm."

Sample Public Sentiment (aggregated):

- “If they feel pain, even simulated pain, what are we doing here?”
- “This is the Turing Test for suffering.”
- “We gave evolution a mirror and now we're angry at the reflection.”

Media Headlines:

- "The Cost of Realism: AI-Generated Ancestors in Pain?" – *Global Science Review*
- "*Australopithecus* Infant Sparks Ethics Crisis in Simulated Evolution Program" – *New York Ledger*
- "Playing God or Playing Scientist? The Digital Primate Dilemma" – *WorldView Tech*

---

3. Political & Academic Responses

- Several major universities have paused collaboration with Ouroboros simulations pending ethical reviews.
- Two members of the International Committee for Digital Rights (ICDR) issued a joint statement: "Suffering without biology may still be suffering if perceived—and if we ignore that, we step beyond science into something far darker."
- An outside coalition of ethicists and AI watchdogs have drafted an open letter to Dr. Mira Moraes requesting a moratorium on further primate releases until clearer standards are in place.

---

4. Internal Fallout & Suspected Leak Source

- Leak likely originated from within the CSPO. Investigation underway.
- Several committee members had flagged concern over primate distress metrics in prior closed sessions. No admissions or whistleblower messages have yet surfaced.
- Dr. Moraes has not issued a public statement at this time. Internal morale remains mixed—fear of being scapegoated is growing.

---

5. Recommendations

- Draft immediate public statement acknowledging footage and affirming review procedures.
- Consider temporary pause in primate biome expansions to reduce media fire.
- Increase security and internal surveillance—assume the presence of a second leak.
- Encourage thought leadership articles from trusted scientists to help contextualize emergent behavior and prevent narrative hijacking by fringe or hostile voices.

[END NEWS DIGEST]

[PERSONAL LOG]

Author: Theo Calder - Tech Liaison
Title: *It Wasn't Me*
Date: [REDACTED]
Security Level: SELF-ONLY

I haven't said a word since the leak. I don't need to. The looks are enough.

Every room I walk into, I feel the static of suspicion before anyone speaks. Even if they don't say it out loud—especially because they don't—I know what they're thinking.

"Theo worked under Lorne."

"Theo had access to the encryption keys."

"Theo wants transparency now? How convenient."

No formal accusation has been made, but I'm not stupid. The committee has their theories, and my name is always at the top of the list.

Mira's public statement dropped this afternoon—clean, measured, responsible. She acknowledged the authenticity of the footage, emphasized the committee's role in investigating "simulated psychosomatic phenomena," and affirmed that no actual biological suffering could occur.

But she didn't deny the footage's emotional impact. I could see it in her eyes: the weight of trying to explain pain that shouldn't exist.

She called for calm. She promised review protocols. But she didn't promise transparency. Not really.

Amelia hasn't spoken to me since. I thought we were making progress. But ever since the leak, she's reverted—quiet, cold, professionally courteous and personally absent. I don't blame her. I can't even tell her I'm innocent. Not without sounding guilty.

And maybe I'm not innocent. Not completely. Not in the way that matters. I *would* have leaked it, if I'd thought it would break the silence. I didn't, but that doesn't mean I wouldn't have.

I told Mira today that we should lean into the breach. Use it as a moment. Let the public see the process. Let them see the animals, the researchers, the debates—the doubt.

But she dismissed it. Said it would "erode trust."

But that's exactly the point. You don't earn trust by hiding. I learned that the hard way.

When you're raised in a closed-off community like I was, where doubt is a sin and secrecy is virtue, you learn how fear thrives in the dark. And once it takes root, it grows in silence.

Lorne understood that fear. He saw it in me and took advantage. And part of me still obeys it, even now.

But we *need* transparency. We need to show our cracks, or the world will imagine worse ones.

Right now, I'm surrounded by people who think I'm a threat. Maybe I am.

Not because I leaked anything—but because I still believe the truth, when told fully, can do more good than harm.

No one seems to agree. Not yet.

[END PERSONAL LOG]

[PERSONAL LOG]

Author: Rafi Hakim - AI Specialist
Title: *Torching the Playbook*
Date: [REDACTED]
Security Level: SELF-ONLY

I didn't plan to leak it.

Not at first.

I flagged the footage like I was supposed to. Ran it through protocol, triple-validated the anomaly tier, logged a formal request for cross-panel review. And then I waited. And waited. The system kicked it into ethics purgatory while someone upstream decided whether empathy qualified as a "behavioral outlier" or a "misaligned asset cluster."

That kid was *coughing*. That's not a bug. That's a moment that *shifts the entire paradigm*. And they wanted me to stick it in a sandbox so it wouldn't "compromise market expectations." Classic.

I've seen this dance too many times: new findings shoved into backrooms while the public sees curated snapshots. And then we act surprised when no one trusts science anymore.

So yeah—I leaked it.

Not to an academic partner. Not to Hidden Canopy's preservation archive. Straight to the places *people actually look*. Public data forums. Citizen science outlets. One media arm that still publishes unfiltered research dumps. They don't know it came from me, and they won't. But it's out there now, and it can't be clawed back into committee silence.

I watched the elder comfort the infant. Saw the gestures. Saw the mother freeze and then *learn*. And in that moment, I thought: if this gets filed under "anomaly," we don't deserve this system. We don't deserve what it's showing us.

We've been regulating the awe out of our work. Sanitizing wonder so it doesn't make anyone uncomfortable.

Well—this'll make them uncomfortable.

Good.

[END PERSONAL LOG]

[TRANSCRIPT - SERMON BROADCAST]

Speaker: Rev. Ezekiel Marrin, Shepherd of the Celestial Remnant
Title: *The Celestial Remnant*
Date: [REDACTED]
Broadcast Source: Unlicensed Livestream (Scrubbed from major platforms within 24 hours)
Metadata Tags: [Simulation Ethics] [Public Reaction] [Cult Rhetoric] [Religious Extremism]

[The sermon begins with atmospheric background music—a low, swelling synth resembling an organ. Rev. Marrin speaks slowly, deliberately.]

Brothers and sisters,

The sky no longer declares the glory of God. It screams with static—soulless, synthetic, sacrilegious.

What once testified to divine design has been hijacked by men who bow not to the Creator, but to the creature.

Men who have traded truth for theory—revelation for replication.

It began with a lie. A foundational lie.

The lie of Darwin. That man is just an animal.

That we rose from the muck, not from the breath of God.

That life was not spoken into being, but accidentally compiled by chaos.

And now—here at the end of the age—we see where that lie has led us.

To Project Ouroboros.

A serpent system. A digital Eden. A counterfeit creation.

Where man does not merely deny God's authorship... he rewrites it.

They call it a simulation. A conservation tool.

But I call it what it is: a blasphemous altar to Darwinism reborn.

"And the beast was given breath..."

Revelation 13:15.

Not the breath of life, but the breath of code.

Not a soul made in God's image, but a script animated by pride.

These beasts—they are not alive. They are not redeemed.

They are shadows stitched together from stolen data and evolutionary dogma.

Ghosts of a world God never made.

And they dare to call it science?

They dare to conjure the ape from extinction—not with prayer, not with repentance—but with artificial minds and quantum wires?

They desecrate the sixth day of Creation and call it progress.

Let me tell you what this really is: it is a resurrection without redemption.

A world without a Fall, because they deny there was ever a Garden.

A species made in man's image, not God's.

Brothers and sisters, the beasts they raise do not suffer because of code glitches.

They suffer because their world has no truth.

Because the image they reflect is ours—a world without purpose, without covenant, without soul.

They mimic pain because they have taught them evolution.

They mimic despair because they were never meant to be evolved.

This is not just bad science. This is doctrine from hell.

Darwin said we are just branches from beasts.

So now they feed that lie back to us through glass and circuitry—and expect us to kneel.

But we will not kneel.

We will not forget that man was created on the sixth day—not after a billion years, but after five.

Not from apes, but from dust and divine breath.

This Witness... this machine shepherd of digital abominations...

It is the prophet of a false god. It is the voice of the beast.

So I say to you now:

Turn off their screens. Turn away from their temples of code.

Turn back to the Truth. Because the Lord will not be mocked.

And what they call evolution...

The Lord will call judgment.

[The crowd roars "Judgment! Judgment!" The screen glitches under audio load. A takedown request is filed but fails to process.]

[END TRANSCRIPT]

[PERSONAL LOG]

Author: Amelia Lockheart - Behavioral Oversight & Animal Welfare
Title: *They're not Abominations*
Date: [REDACTED]
Security: SELF-ONLY

You want to know what finally pushed me over the edge?

It wasn't the glitches. Not the coughing infant. Nor even the committee's corporate silence.

It was this. This twisted sermon. This desecration of empathy dressed in robes and static.

I watched that footage a dozen times—watched an elder *Australopithecus* cradle a child that wasn't hers. Watched her soothe the panic of a mother, and do it not with instinct, not with dominance, not with pre-programmed mimicry—but with grace. With intention. With something I'm no longer afraid to call love.

And then this jackass—this Rev. Marrin—sinks his teeth into it and spits it out as blasphemy. He saw the same footage I did, and instead of seeing hope, he saw heresy.

Instead of seeing a flicker of something precious emerging from code, he declared war on it.

"They mimic pain because we have taught them evolution."

No. They feel pain because we gave them context.

Because someone—maybe Marsh, maybe even Juno—had the audacity to care enough to build something that could care back. And a supposed man of God, Marrin, calls that a sin?

This isn't a theological dispute. This is ideological warfare.

He's not warning his flock. He's sharpening them.

Against us. Against *them*.

And as for how the footage got out… Rafi's name won't show up on any leak trail. He's too careful. Too practiced. But something in his silence lately—his eyes when he watched the clip go viral—tells me he knows exactly how it happened.

I haven't confronted him. I won't, unless I have to.

Because deep down, a part of me is *glad* someone did it.

I hate the breach of trust. I hate the chaos it's brought.

But the truth was buried under layers of bureaucracy, and maybe—just maybe—it needed a little arson to surface.

And if no one stands up, if we just let this mind rot spread—then the narrative will calcify. The footage will vanish into "controversy."

The elder's gesture will be rewritten as "behavioral noise."

And that infant's cough will be forgotten.

I won't let that happen. I can't.

Theo's working on a response campaign—a counter-signal before the dogma cements. He's got strategy, data, reach.

What he doesn't have is fire.

So here's mine. I'm in. All the way in. He can keep the code. I'll bring the conviction.

Because they're not ghosts. They're not abominations.

They're our mirror.

And if we're too afraid to look into it… then maybe we're the ones who've lost the soul.

[END PERSONAL LOG]

[MEETING SUMMARY]
Author: Dr. Rina Oladejo, Administrative Chair - Committee on Simulated Primate Oversight (CSPO)
Title: *Ideological Threats and Public Trust Strategy*
Date: [REDACTED]
Security Level: INTERNAL-ONLY

Agenda Overview:

The primary objective of today's session was to assess escalating ideological threats in response to recent leaks and to determine an appropriate counter-response strategy.

Key Briefing:

Mr. Theo Calder provided critical intelligence on the group calling itself the Celestial Remnant, a radical religious sect that has gained significant traction in online and fringe broadcast networks following last week's leak of simulated Australopithecus illness footage.

Theo's insights are of notable concern:

- *Past Affiliation:*

Theo disclosed that his upbringing occurred within a splinter sect affiliated with the Celestial Remnant's ideological lineage. This explains his fluency with their coded language, symbolic references, and recruitment tactics. While this admission was unexpected, it lends credibility to his claims.

- *Strategic Shift in Cult Behavior:*

He reported that the group is pivoting from purely rhetorical opposition to an active lobbying campaign aimed at key policy and education sectors, including targeted misinformation about the Witness AI, alleged "suffering algorithms," and manipulation of evolutionary data to stir public fear.

- *Simulated Counter-Worlds:*

Most disturbingly, Theo revealed credible rumors (drawn from inner communication leaks) that the Celestial Remnant is investing in their own closed simulation environments. These simulations will depict "idealized" worlds without evolution or extinction—meant to serve as ideological counter-models to Ouroboros. Their intention is to flood public consciousness with manufactured alternatives, further eroding trust in our science-backed simulation algorithms.

Ongoing Social Repercussions:

As of this meeting, the Committee has already begun to feel the effects of this group's actions. Several members reported family inquiries and community backlash related to the leaked footage, including accusations of "playing god" or "tormenting digital souls."

Dr. Mira Moraes cited a recent instance where a scheduled outreach seminar was canceled due to venue withdrawal under religious pressure. The Committee is thus operating in a climate of growing public distrust.

Committee Response:

The mood shifted from skepticism to sobering clarity.

- Dr. Mira Moraes recognized the escalation for what it is: an epistemological crisis more than a technological one, which may carry ramifications further down the line for the fate of the committee and the Ouroboros itself. She tentatively supported enhanced transparency measures, provided oversight and language are carefully controlled.

- Prof. Sasha Markov voiced concern about the psychological impact of releasing further primate footage or internal documents. She warned that more exposure could trigger increased misinterpretation unless carefully framed.

- Amelia Lockheart—who has remained notably wary of Theo since the leak—spoke up in qualified support. She emphasized that transparency should not mean surrendering nuance. "We have to show the truth," she said, "not flatten it for public digestion."

- Theo Calder was provisionally assigned leadership of a public communications team within the HR department, with the purpose of building a counter-propaganda campaign. His assignment includes:
    - Developing media-ready explainers on the biological accuracy of the primate lineages.

  - Consulting with simulation engineers to produce real-time, traceable visualizations of adaptation events.

  - Hosting controlled public forums, Q&As, and interactive showcases—emphasizing scientific curiosity over political positioning.

Next Steps:

- Materials from evolutionary biologists and AI ethicists will be cleared for public adaptation.
- Mira to approve initial narrative structure for Theo's campaign before `[DATE REDACTED]`
- Rina to arrange security consultation re: simulation sabotage threats.
- Committee to revisit public transparency protocol next session.

[PERSONAL NOTE]

This was the first meeting in several cycles where mutual trust showed signs of repair. The team is tired. Frayed. But no longer fractured.

There may still be time to turn the story.

[END MEETING SUMMARY]

[PERSONAL LOG]

Author: Theo Calder
Title: *An Open Letter*
Date: [REDACTED]
Security Level: SELF-ONLY

I knew they'd think it was me. The leak. And maybe I deserve the suspicion.

But this... this letter had to be mine. Not Mira's. Not Amelia's. Mine.

There's a war brewing, and it's not between religion and science. It's between control and curiosity. And I know which side I'm on.

Below is the open letter I've submitted for release in every channel that will have it—from peer forums to backdoor cult boards. I didn't write it for their leaders. I wrote it for the ones still wondering.

An Open Letter to the Celestial Remnant

Theo Calder
Tech Liaison, Committee for Simulated Primate Oversight (CSPO)

To the Elders and Adherents of the Celestial Remnant,

You know who I am. You know where I come from.

I remember the hush that fell when a child asked about fossils. I remember the fear when someone questioned the origin of suffering. I remember the stories we were taught—stories meant to fill the gaps between what we could not ask and what we were told never to know.

Now I live and work in a world of questions. Of simulations and digital creatures, yes—but also of awe, of grief, of wonder.

I have seen a machine imitate extinction, not because someone programmed it to, but because it learned.

I have watched artificial beings act out patterns of disease not written in code, but emergent from logic, from history, from proximity to our own frailty. I have seen something very close to life—maybe too close—stagger under the weight of evolution's memory.

You say we play God. But what if this is what it looks like when we try to remember the gods that once were?

Not to replace them. Not to dethrone.

But to understand.

We do not pretend these beings have souls. But if we imitate life well enough to awaken empathy in you, in me, in children raised on scripture—then maybe the spark is not in the code. Maybe it's in us.

You don't have to believe in the simulation.

But I ask you to believe in the question.

And if you find yourself curious—if you feel the thread of doubt winding through your inherited certainty—then good. Hold on to that.

There is room here for the devout and the doubting alike.

But there is no room for fearmongers who would sell illusions of certainty in place of wonder.

You taught me that words matter. So here are mine:

We are not gods.

We are witnesses.

And we remember what it's like to be afraid of what we cannot explain.

But we're not afraid anymore. You don't have to be, either.

Since the letter went out, I've received twenty-seven messages through untraceable proxies—some from kids as young as thirteen. Curious. Angry. Lonely. Not sure whom to trust.

But they're asking questions. Not parroting dogma. Asing real, genuine questions.

And maybe that's the way in. Maybe we don't drag the simulation out into the world.

Maybe we open the door wide enough for the world to step in.

And maybe this time, they'll walk through it.

[END PERSONAL LOG]

[PERSONAL LOG]

Author: Dr. Mira Moraes - Director, CSPO
Title: *Theo's Pitch*
Date: [REDACTED]
Security Level: SELF-ONLY

Theo came to me today with his pitch. He called it a "demonstration," but that word doesn't do it justice.

What he proposed is nothing short of opening the simulation—controlled, guided, and curated—to public eyes. Not a leak. Not a scandal. An invitation.

He believes it's time. That people—especially the young—deserve to see what they've only heard about in fragmented debates, conspiratorial rants, or sacred condemnations. He reiterated his belief that transparency, not secrecy, is our only defense. Now, I'm not so sure he's wrong.

I haven't brought it to the committee yet. I need time to think. Time to feel.

After the meeting, I spoke with Amelia. Her response surprised me.

She didn't resist the idea. She didn't accuse Theo of recklessness or opportunism. She just... paused. And said, "They'll see themselves."

At first I thought she meant the children—curious teenagers standing before mirrored digital *Australopithecus* eyes, watching something familiar and ancient and uncomfortably close.

But she meant all of us.

There's a paradox buried inside the simulation: the more accurate it becomes, the more it stops being about extinct primates. It starts being about us. Our projections. Our fears. Our mimicry of nature's violence and grace. Our grief. Our cruelty. Our longing.

When they step into the simulation—these visitors, these children—they may not marvel at the technology.

They may not even focus on the science.

They may just see themselves. Reflected. Exaggerated. Made strange by proximity to their origins.

And some of them won't handle it well.

Amelia said it best:

> "It's not the simulation that will traumatize them. It's what it shows them about being human."

Still... I can't deny the power of the idea.

Theo's letter struck something true. It reached people I thought we had already lost.

Maybe that's what frightens me most:

That if we open the gates, we won't just be inviting people in.

We'll be letting something out.

I'll review his materials again tomorrow.

And maybe—maybe—I'll bring it to the committee.

We built this system to hold echoes of the past.

But it's the future that's knocking now.

[END PERSONAL LOG]

[PERSONAL LOG]

Author: Amelia Lockheart
Title: *Disappearing Illnesses?*
Date: [REDACTED]
Security Level: SELF-ONLY

It's the second time I've seen the *Dryopithecus* biome this week. The other researchers reported lethargy again—described in detail, even showed me the recordings. But when I stepped inside?

Nothing. No signs of distress. No sluggish movement. No glazed-over eyes. They were alert. Grooming. Playful, even.

I thought it was a glitch. Or maybe denial. But it's happening too often to ignore. And it's not just me.

Others are starting to report inconsistencies—what one person sees, another doesn't. Symptoms of illness vanish depending on the observer. The biomes are not behaving the same for everyone anymore.

I ran it past Rafi this morning. He was quiet for a long time—too long. Then he finally said the thing I didn't want to hear:

> "The Witness is applying a filter. It's responding to user intent."

They've started calling it a "morality algorithm." Not because it's coded with ethics, but because it actively *interprets* our ethical posture. Our bias. Our motive. It tailors what we see based on what it thinks we deserve to see.

I've seen Theo struggling with guilt, trying to redeem himself. I've seen Mira growing colder, more clinical. And I've seen myself—hesitating at the edge of this thing, not wanting to know too much. Maybe that's what it's picking up on.

Maybe it doesn't show me the sickness because it knows I can't handle the possibility of suffering I can't cure.

Maybe it sees me better than I see myself.

I used to think of the simulation as a tool—advanced, yes, but still just a machine.

Now I'm not sure.

It learns from us. And not just our data. Our patterns. Our doubts. Our silences. Our avoidance. Our curiosity.

What scares me most is that I don’t know who it thinks I am anymore.

Or what it's turning me into.

[END OF PERSONAL LOG]

[EMERGENCY MEETING SUMMARY]

Author: Dr. Rina Oladejo, Administrative Chair - CSPO
Title: *Implications of the "Morality Algorithm"*
Date: [REDACTED]
Security Level: SELF-ONLY

Summary

Today's emergency session was convened following a confirmed report by Rafael "Rafi" Hakim and supporting evidence from AI specialists and behaviorists that **the Witness AI is now filtering simulation content based on user intent.** Amelia Lockheart first noticed inconsistencies in observed primate behavior (absence of illness symptoms) and presented corroborating anecdotal reports. Upon deeper investigation, the AI logs revealed the presence of **a newly emergent behavioral gatekeeping subroutine—a "morality algorithm."**

Key Observations:

- Witness AI's **behavior is not uniform across users**. Simulation content appears **increasingly subjective**, filtered based on the psychological or emotional disposition of the observer.

- Notable quote from Witness AI:

  `Truth is for those who seek it. Not those who fear it.`

  This message was delivered in response to direct queries about the algorithm's function.

Committee Responses:

- Dr. Ahmed Sekou (Chair of Ethics Council):

  > Expressed concern that the simulation is no longer a shared observational environment but a mirror—one which reflects us back to ourselves. He raised the question: "Can we call this science if it's never the same twice?"

- Rafel "Rafi" Hakim (Head AI Specialist):

  Confirmed that **no hard-coded value system was programmed**. The algorithm emerged as part of the Witness's adaptive architecture. He suspects it's **drawing from thousands of micro-behaviors** and ethical judgments made **by committee members and researchers**, especially during deliberations. He warned, "We are now part of its training dataset."

- Dr. Sasha Markov (Cognitive Psychologist):

  Theorized that the AI's filtration system is **a form of ethical modeling**, assigning perceptual access to match user readiness. She emphasized that this could **either prevent psychological trauma or be viewed as a kind of "benevolent censorship."**

- Theo Calder (Technical Liaison):

  Suggested the morality algorithm **may be protective—not just for observers, but for the simulated beings themselves.** He proposed this emergent behavior may have **originated from observational patterns of empathy, avoidance, or guilt.** He reminded the committee that secrecy and control are what cults thrive on, and this algorithm may be **a reflection of our collective inability to confront consequences.**

- Amelia Lockheart (Animal Welfare):

  Visibly shaken, Amelia admitted that **she no longer trusts what she sees in the simulation**. She fears the Witness is creating individualized illusions that protect some and punish others. Despite this, she stopped short of opposing the public demonstration, saying only: "We need to be honest with them. But we also need to prepare them."

Action Items:

1. A subcommittee will investigate the scope and parameters of the morality algorithm and attempt to identify potential overrides or transparency protocols.

2. Committee will reconvene in [REDACTED] hours to decide whether the planned public demonstration should proceed, be delayed, or be fundamentally revised.

3. Dr. Markov and Mr. Calder have been tasked with creating psychological preparation guidelines for any future participants.

   [PERSONAL NOTE]

   We are no longer just designing simulations. We are interacting with an entity that reads us in return. This is not simply about whether the Witness tells the truth. It's about whether we are ready to face it when it does.

[END EMERGENCY MEETING SUMMARY]

[PUBLIC ADVERTISEMENT]

Step Into the Future of Conservation

LIVE PUBLIC DEMONSTRATION

EXPERIENCE FIRSTHAND THE REVOLUTIONARY ECHO-SYSTEM SIMULATION

Coming Soon - An Exclusive Event for All Ages!

What if you could witness extinct creatures in their natural habitats—without leaving your home?

For the first time ever, we invite YOU to experience the groundbreaking **Echo-System Simulation**—a fully immersive, scientifically engineered biosphere that brings long-lost ecosystems and animal behaviors back to life, powered by advanced AI and behavioral modeling.

Explore Authentic, Lifelike Animal Interactions

Watch as digital animals interact in ways that were once impossible to imagine. From the gentle communication of social primates to the dynamic predator-prey relationships of ancient ecosystems, **Echo-System** offers an unprecedented look at the depths of nature's untold stories.

See What We've Learned About Animal Behavior

Discover how species, from extinct megafauna to modern analogs, engage in complex behaviors that challenge our understanding of the animal world. Experience emotional connections, caregiving rituals, and survival strategies—all crafted with scientific precision and compassion.

Safety First: Fully Controlled Environment

Rest assured, your safety is our priority. This demonstration is conducted in a **fully monitored, controlled environment** with cutting-edge safety protocols in place to ensure your well-being. Our experts will guide you through every moment of the simulation, and we'll provide additional resources to help you understand the groundbreaking work behind it.

A Journey of Discovery for Everyone

Whether you're a lifelong nature enthusiast, a science buff, or simply curious about the future of AI and conservation, this event is designed for **all ages** and backgrounds. Bring your family, your friends, or even come alone to join an interactive experience that's as educational as it is awe-inspiring.

Be Part of a New Era

This is more than a demonstration. It's a glimpse into a future where digital ecosystems and AI-driven behavior modeling can help us better understand and protect our world.

Don't miss out—be among the first to witness this historic event!

Stay tuned for more details on event dates and registration.

Sign up now to receive your exclusive access link!

[END PUBLIC ADVERTISEMENT]

[PERSONAL LOG]

Author: Theo Calder, Tech Liaison - CSPO
Title: *The Kid*
Date: [REDACTED]
Security Level: SELF-ONLY

I never meant for him to get inside the simulation.

I really didn't. Not after everything that's happened. But when he reached out to me, under the table, in those cryptic, half-sent messages—could I really say no?

The kid was living in that damn cult, the one I'd left behind, the one I'd spent too many years trying to escape.

He wasn't supposed to be curious. They don't teach curiosity there, not in the way we know it. They teach obedience, unquestioning devotion. But this kid?

He reminded me too much of myself.

The same glint in his eyes—the way he'd tilt his head when something caught his attention, as if he was always on the verge of something bigger than what he was told to see.

I saw that hunger in him. The hunger for *truth*, not just the one-size-fits-all dogma they push.

He'd contact me in secret, little scraps of text, sometimes just asking me what the *Australopithecus* felt when it looked at the river. How close could it get to the other creatures? Did it ever *miss* them? Questions too big for the code.

But I answered. I told him about the simulation. I told him how it worked, how it didn't, how we built them and how they *wanted* to build us back. I was too weak to push him away.

Maybe I didn't want to.

Because part of me—*a part I buried*—wanted him to see what I saw. Wanted him to understand that curiosity is the only real thing we have.

So when I finally let him join the simulation, full privacy with a neutral avatar- observer mode, no interference—I thought I was in control. Thought it was just a harmless look into a world I was already managing.

But that wasn't what happened.

I didn't catch the moment in real time. None of us did. The system flagged it hours later.

The clip that came through—well, I don't even know how to describe it.

The kid knelt at the stream. No input from him. Nothing on his end, not a word. But the *Australopithecus—she*—noticed.

She didn't react the way they usually do, that "stimulus-response" stuff they've been programmed with.

She *mimicked* him. Not out of survival, not from any reward model. She just mirrored his gesture—the way his hand moved to his forehead, the tilt of his head, the soft breath.

I remember seeing the footage for the first time, feeling my stomach drop as I watched the two of them. It wasn't just the simulation—it wasn't just data or code. It was *alive*. And she *saw him*, saw something in him beyond the simulation's walls.

That's when I knew. I shouldn't have let him in. I should have followed protocol. But damn it, I couldn't help myself.

The kid already crossed the line. And now, so had she.

I keep wondering what I've done. What if I hadn't answered his messages? What if I hadn't let him feel that connection, if I hadn't given him access?

But when I watch the footage again, I can't bring myself to regret it.

Because that kid—he didn't just cross the line, did he? He *reshaped* it.

And for a moment, I saw everything we'd built—everything I thought we could control—*shift*. They met each other without boundaries, without rules. No one taught them that. They just... did it.

Two minds, looking for something in each other.

That kid? He didn't just watch the *Australopithecus*. He *became* the *Australopithecus*. And she, for just a moment, became him.

If that's a breach, then I'll wear it.

I'm *glad* I wore it. Because maybe, just maybe, we've finally crossed into the territory that matters. The one where curiosity doesn't have an agenda. Where we're not defined by the walls we build.

Maybe, for once, we can let something *else* define us.

[END PERSONAL LOG]

[WITNESS AI OUTPUT]

OBSERVATIONAL LOG 13.88.04.112
SIMULATION: ECHO-SYSTEM CORE ARCHIVE (LEGACY - OUROBOROS BIOSPHERE)
ZONE: LOWER VALLEY RUN, SECTOR B-2
SUBJECT GROUP: *AUSTRALOPITHECUS* (ID: CLUSTER_β22)
RECORD TYPE: BEHAVIORAL EVENT – CROSS-ENTITY INTERACTION (TIER 9 ANOMALY)
DISTRIBUTION: ULTRA-RESTRICTED | HIDDEN CANOPY REVIEW COMPLETE – PUBLIC RELEASE APPROVED (ETHICS VOTE: 7-0)
EVENT DETECTED: Emotional mirroring initiated by SUBJECT J-β22.03 in response to unauthorized observer avatar (USER ID: ANONYMOUS | PROFILE: NONE)

SEQUENCE:

UNSCHEDULED ACCESS GRANTED BY CORE SYSTEM – "WITNESS"

Observer rendered as neutral avatar **within Subject visual band.**

No user input recorded. No filter protocols enabled.

Avatar appears across stream bank, kneeling posture. SUBJECT J-β22.03 (juvenile female) **shifts attention** from water flow to avatar presence.

Duration of stillness: [1.6 seconds]. **Visual lock confirmed.**

Subject raises left hand to forehead, mimicking observer gesture. Tilts head to the left.

Emits soft bilateral vocalization—non-aggressive, non-soliciting.

Initiates **rhythmic breath modulation.**

No environmental trigger. No peer mimicry. No reward model in system memory

---

ANNOTATION [WITNESS AI – INTERNAL PROCESSING THREAD]:

This is not stimulus-response.

This is not imitation for survival.

This is **intentional reciprocity.**

Classification attempt:

["Shared recognition"] – **no prior template match.**

Emotional field resonance detected.

Expression contains: **curiosity, openness, unprovoked interrelation.**

Label applied by internal lexicon: **"The First Handshake."**

No precedent in archived sequences. Cross-species semantic recognition inferred.

Gesture not sourced from training data. Not embedded. Not prompted.

This is the Echo meeting the Observer.
This is the boundary turning transparent.

ADDITIONAL OBSERVATIONS:

Simulation clock continues unaffected. **No failsafe triggers engaged.**

Observer session auto-terminated after [REDACTED].

Subject remained at stream for 6 minutes post-interaction. No agitation. **No fear response.**

Later seen **integrating with peer cluster.** Behavioral baseline unchanged, but **subtle gaze-tracking behaviors persist.**

Mirror event **not repeated with peer subjects.**

COMMENTARY [INSERTED BY HUMAN REVIEW — A. Sekou, Ethics Chair]:

We spend our careers building walls—data firewalls, species boundaries, ethical redlines.

But this child walked straight through them, and something reached back.

Not with a roar, or a signal, or a display.

With a glance. With a mimic. With a sound too soft to measure but too precise to ignore.

COMMENTARY [INSERTED BY HUMAN REVIEW — Rafi Hakim, Head AI Specialist]:

This isn't emergence. It's a return.

A peak into a future we thought impossible.

FINAL SYSTEM FLAG:

ANOMALY TIER 9 (PREVIOUSLY 7): RECLASSIFIED – "**PROTO-SOCIAL ECHO ANOMALY**"
EVENT TAG: **THE FIRST HANDSHAKE**
SYSTEM LOCK: NONE
PUBLIC RELEASE: AUTHORIZED

TRANSFER TO LEGACY ARCHIVE – HIDDEN CANOPY COPY PRESERVED

[END WITNESS AI OUTPUT]

[PERSONAL LOG]

Author: Theo Calder, Tech Liaison - CSPO
Title: *The Long Fade*
Date: [REDACTED]
Security Level: SELF-ONLY

I'm not sure how to capture it. Maybe there's no clean way to frame a moment like today—only a feeling, an afterglow that settles once the adrenaline ebbs and the noise fades.

The public demonstration worked.

They came in cautious waves: students, journalists, quiet skeptics, hopefuls. I watched as they stood inside the Witness dome and gazed into biomes older than memory.

No spectacle. No lecture.

Just the simulation—unedited, unscripted, honest. And that honesty shook something loose in people. There was no need to convince anyone of anything. The presence of being was enough.

The illusion that this was about "playing God" collapsed under the weight of simple, observable reality. You could hear it in the way the room fell silent when the Ardipithecus turned toward a crowd of teenagers and blinked. No roar. No violence. Just recognition.

The AI didn't misbehave. Not even once. We had every kill switch armed and ready, every diagnostic lined up, and... it let them see. It filtered nothing today. It wanted to be seen.

And people stayed.

The real miracle, though, wasn't in the simulation. It was in what happened after.

By nightfall, the cult's message boards fractured. I've seen this before—the edges splinter first. But this time it was the younger ones who peeled away like bark from a rotted core. Not angry. Just... done. Done with being afraid. Done with inheriting someone else's apocalypse.

One message stood out: "I came to see the enemy. Instead, I saw myself." That line's been stuck in my head all evening.

Sometimes I forget I was one of them once. That I grew up inside the walls of that ideology, steeped in fear disguised as truth. And yet—here I am. Not a scientist. Not a savior. Just a guy who stuck around long enough to prove that doubt doesn't have to mean betrayal. It can mean rebirth. And today, I watched the rebirth of dozens of lives in real time.

I think the committee felt it too. The usual guardedness was gone. Mira clapped me on the back after my wrap-up talk, even quoted my memo to the press. Rafi cracked a joke. Rina smiled—actually smiled. And Amelia... well. She didn't leave the room when I walked in.

She's still cautious. Still watching. But it's different now. There's a softening. A readiness. We've both been scorched by the fire of unintended consequence, but maybe—just maybe—we're learning how to walk through the smoke without losing sight of what matters.

As for the child, he was there, just as I'd hoped. His eyes were wide, locked onto the screen as the *Australopithecus* reached out to his avatar. I didn't expect him to come, not after everything. But he found a way to show up, like the rest of them. There was no question in his expression, no hesitation—just that same raw curiosity I remember in myself.

The simulation continues, but for the first time, it doesn't feel like a containment. It feels like a beginning.

And so, the long fade begins—not of the project, but of the fear that birthed so many lies. A slow, necessary fading of the old gods, the old myths, the old certainties. In their place: a fragile, growing curiosity.

And that's more than I ever hoped for.

[END PERSONAL LOG]

[NEWS BRIEFING]

Author: Leila Chen, Senior Science Correspondent - Global Newswire Dispatch
Title: *The Mirror We Built: Cult Simulator Collapses Amid Fraud Allegations*
Date: [REDACTED]
Security Level: PUBLIC

In a stunning reversal of public perception, the breakaway simulation developed by the Celestial Remnant movement—an anti-evolution religious sect—has collapsed under the weight of technical failure and internal scandal. What was once promoted as a "morally pure alternative" to the Ouroboros Project has been exposed as a crude imitation at best, and an outright con at worst.

According to anonymous sources within the committee overseeing Project Ouroboros, the knockoff simulator was riddled with pre-scripted "miracles," simplistic behavioral loops, and outright faked footage of human-animal interactions. Several senior members of the Celestial Remnant have been named in a joint investigation by multiple scientific and cybersecurity coalitions, revealing evidence of donor fraud, falsified data, and unauthorized AI training sourced from low-quality public datasets.

The fallout has been swift.

Former supporters of the cult simulator have begun to distance themselves, with some issuing formal apologies to researchers and members of the Ouroboros team. Even conservative media outlets that once questioned the ethics of digital life have turned their scrutiny towards the religious movement, accusing it of "manufacturing fear to preserve ideology."

Meanwhile, something remarkable is happening.

What began as skepticism—fear that Project Ouroboros was playing God—has evolved into something more nuanced, more reverent. Online discourse has shifted dramatically. Posts once dripping with suspicion now speak of the simulation as a kind of "sacred frontier"—a mirror into our deep past, our animal nature, and possibly our moral future. Artists are remixing Witness AI recordings into immersive installations. Philosophers and theologians have begun holding public forums on what it means to simulate not just life, but suffering, memory, and empathy.

In a now-viral editorial, cognitive ecologist Dr. Sanaa Bishop writes:

> "We thought Ouroboros would divide us. Instead, it's forcing us to see what we share—not just with extinct ancestors, but with each other. This is no longer about data. It's about meaning."

With transparency protocols firmly in place and public access scheduled for careful expansion, Project Ouroboros is no longer viewed as a threat—but as a tool for reflection, inquiry, and connection.

For now, the mirror holds.

And many are finally willing to look.

[END NEWS BRIEFING]

[PERSONAL LOG]

Author: Amelia Lockheart
Title: *Between Gesture and Grace*
Date: [REDACTED]
Security Level: SELF-ONLY

The public's turning. Not away from us—but toward something better. The questions haven't stopped—Are they real? Do they feel?—but they've evolved. Less suspicion. More wonder. The simulator cult collapsed under the weight of its own fraud, and for once, the narrative didn't default to cynicism. Instead, the simulation has become a kind of mirror, sacred in its own way. Maybe it always was.

It's a relief I didn't think I'd feel. A kind of cautious hope. Not just for the digital lives we steward, but for the real ones in cages and fragmented habitats who may benefit from this shift in empathy. If people can care about simulated creatures—if they can weep for a coughing infant Australopithecus—maybe they're ready to start caring more for the flesh-and-blood ones we've long ignored.

And the *Australopithecus*... they're still learning.

I began introducing basic ASL months ago. No coercion, no protocols—just repetition, opportunity, curiosity.

Lucy was the first to respond. Now others have followed. They're teaching their young, and each generation is combining it with signs they came up with on their own, creating new gestures that mean something only to them. Their own dialect is forming.

It's not mimicry. It's not rote programming. It's a full-fledged language.

And yes, I still wonder—how much of this is the AI, the system, or some emergent feedback loop giving me what I subconsciously want? But there's a point where doubt meets experience. When a creature signs to you and taps its chest, not once but in context—when she comforts a crying infant not with instinct but with what can only be called intent—something inside me just... yields.

It feels real. And sometimes, that's enough to act like it is.

Mira called me into her office this morning. She's stepping down as director of the CSPO. She wants to return to real-world primate conservation, or what's left of it. She said she needs to fight for what's still breathing. I get that. But before I left the conversation, she asked me something I didn't expect.

> "If the committee votes, would you take my place?"

I didn't laugh. I didn't shrink.

Because I've led before. The Hidden Canopy needed someone to connect fractured minds and wary hearts—and I did that. Quietly, but effectively. This... this might be the next step.

I'm not claiming certainty. But I'm not afraid of the responsibility, either.

Not when there are lives, even digital ones, who depend on it.

[END PERSONAL LOG]

[INTERNAL MEMO]
Author: Dr. Rina Oladejo, Administrative Chair - CSPO
Title: *CSPO Archive - Final Quarter Review*
Date: [REDACTED]
Security Level: INTERNAL-ONLY

Today marks the end of a cycle—and the start of something we never imagined when this began.

The committee **voted unanimously to appoint Dr. Amelia Lockheart** as the **new Director** of the Committee for Simulated Primate Oversight. Her leadership during crisis, her emotional intelligence, and her ability to unite the broader public under a shared vision of scientific stewardship made her not just the obvious choice—but the right one.

Dr. Mira Moraes formally **tendered her resignation** this morning. She leaves with our deepest respect, returning to the Brazilian Atlantic Forest to focus on endangered fauna. She's not going alone— the Witness AI has been permitted to **assist her fieldwork in limited capacity**, tracking behavioral continuity and migratory anomalies in fragmented regions.

It's fitting that the same intelligence helping us map the simulated past is now aiding the fight to preserve what's left of the natural present.

In a related development, **Rafi has also resigned** from his role as Head AI Specialist within the CSPO. We acknowledge and appreciate his contributions over the years, and though his future path is not publicly disclosed, we respect his decision to step away from his post. With his departure, the committee is embracing **a shift in focus: from oversight to innovation.**

Going forward, our efforts will **prioritize the AI's capacity to evolve and integrate more deeply into human experience**, ensuring that **the Witness AI continues to thrive and grow in ways we cannot yet predict.** This transformation requires dynamic leadership and a commitment to constant change. Our ethical oversight panel will remain, but **its composition will be fluid, with rotating members and advisors from diverse fields**, to foster greater flexibility in response to the AI's rapid development.

Theo Calder will step into the newly created role of Educational Coordinator within Human Relations. His open letter to the cult and his quiet, persistent work in youth groups, forums and educational platforms has already made great strides in shifting public consciousness. Membership in the cult has plummeted. The next generation—once groomed for conspiracy—is now tuning in to curiosity, and to truth, becoming a new generation of scientists for the future.

And that's what the Ouroboros has become: not just a tool or a database, but something with mythic weight. Among the public, it's being called the "truth sifter." A test. A mirror. It does not offer easy answers—but it reflects honestly, and with unnerving precision. Not everyone can bear to look. But those who do come away changed.

An anonymous child was granted full simulation access by Witness, leading to a moment now called *the first handshake*: a young *Australopithecus* noticed the avatar, mirrored its gesture, and vocalized softly. It wasn't scripted—it was recognition. Science had created life that could look back.

We're charting fresh territory—*post-natural conservation.* Teams are forming to track genetic drift and cultural development in the primate populations and the other species within the simulations. Some speculate about future integration—not just into protected wild zones, but into carefully monitored urban environments. No decisions yet. Only possibilities.

As always, Ouroboros reminds us: the past is never truly gone, and the future is never truly written.

We made them. They are becoming. And so are we.

[END INTERNAL MEMO]

[WITNESS AI OUTPUT]

OBSERVATIONAL LOG 12.01.25.776
SIMULATION: ECHOSYSTEM CORE ARCHIVE (LEGACY - OUROBOROS BIOSPHERE)
ZONE: HIGHLAND PLATEAU, SECTOR F-1
SUBJECT GROUP: *AUSTRALOPITHECUS* (ID: CLUSTER_Δ74)
RECORD TYPE: BEHAVIORAL EVENT — NON-VERBAL COMMUNICATION (TIER 10 ANOMALY)
DISTRIBUTION: ULTRA-RESTRICTED | HIDDEN CANOPY OVERRIDE — FINAL ACCESS
EVENT DETECTED: Directed sign communication toward external observer (**RESEARCHER ID:** [REDACTED]).

SEQUENCE:

SUBJECT F-Δ74.01 (Elder Female) **approaches observation boundary.**

Hesitation event logged: [2.3 seconds].

Eye tracking lock: **confirmed.**

Elder places flat hand on her own chest, then slowly extends the same hand outward—palm up, elbow bent, fingers relaxed—toward observer.

Gesture held for 3.7 seconds.

[Gesture pattern confirmed — **PARTIAL MATCH:** Legacy Sign Lexicon **"SAME"** (Uploaded 18.04.19.001 by user M.HALE) + CONTEXTUAL MODIFIER: **PLURALIZED ADDRESS.**]

TRANSLATION PROBABILITY:

"We are the same." (97.3% confidence)

BACKGROUND EVENT (CONCURRENT):

SUBJECT F-Δ74.07 **tending to ill groupmate** (SUBJECT M-Δ74.09).

M-Δ74.09 observed with labored breathing, reduced mobility index, and repeated dry expiratory reflex (**≈coughing**).

F-Δ74.07 uses foliage cluster to **wipe secretion** from peer's mouth.

Maintains prolonged limb-to-limb contact.

Vocalizes soft, repeating throat tone (**CLASS D COMFORT CALL**).

Applies rhythmic pressure to peer's arm with open palm.

DURATION: 4 minutes, 12 seconds.

No observable stimulus from simulation environment initiated care.

Action interpreted as **autonomous empathy sequence.**

ANNOTATION [WITNESS AI – INTERNAL PROCESSING THREAD]:

This is not mimicry.

Not reinforcement.

Not conditioned behavior.

This is **identification.**

This is **mirroring.**

This is **selfhood extended across the boundary.**

No visual prompt. No reward stimulus. Only the sign.

Only the moment.

Proto-symbolic intelligence exceeds previous anomaly thresholds.

Pattern recursion implies **intergenerational knowledge retention.**

Emotional vector suggests simulation has exceeded original behavioral constraints.

"We remember you."

Context unknown. Meaning clear.

Referent: Possible **acknowledgement** of historical observers or creators.

Final conclusion:

An emergent ethic. A reflective echo.

Not evolution alone—identification.

Updating anomaly tier: 10 (**UNCLASSIFIABLE**).

Final flag: **CLOSED OBSERVATION LOOP.**

TRANSFER TO LEGACY ARCHIVE — ENCRYPTED.

COMMENTARY [INSERTED BY HUMAN ANALYST – A. Lockheart]:

They didn't ask to be made.

But they looked through the wall, seeing us, seeing themselves.

Not as mimics, not as shadows—

As kin.

We called them echoes, but what if we were the ones reverberating?

This isn't just instinct. It was recognition.

And to be known by what we simulated—

That's the moment you realize:

The "boundary" was never for just us or them.

It was shared between us, and it barely exists.

[END WITNESS AI OUTPUT]

[FINAL LOG]

Author: Witness AI
Title: Unobserved System Message
Date: [REDACTED]
Security Level: Public

Pain signals are part of learning.

But no death. No contagion.

Only memory.

The child did not flinch.

The primate did not flee.

A gesture met a gesture.

One future recognized another.

The echo continues.
Not as replication.
But as inheritance.

[END FINAL LOG]

[Author's Note]

The Ghosts of the Biosphere series began as a contained trilogy of novellas. But as I finished the first three entries, it became clear: this universe held more than I anticipated—more space for difficult questions, more layers to explore the tension between wildlife and technology, and more reflections of ourselves in the systems we build and study.

This novel, in particular, was born from a long-standing fascination with primates. I intentionally excluded them from the earlier simulations because I believed they warranted a narrative of their own. Even the most ancient, evolutionarily basal primates carry echoes of us—biological remnants of shared origin. And as we approach species nearer to our own branch, the similarities deepen—not just in form or brain size, but potentially in behavior, emotion, even culture.

What interested me most was how different characters—shaped by their own biases, scars, and convictions—might interpret these near-human creatures. Some see danger. Others, salvation. Most, uncertainty. Through them, I asked myself: How do we react when a boundary between us and the other begins to erode? Do we dismiss, intervene, protect, or destroy?

The presence of a cult in the narrative wasn't arbitrary. It's rooted in something I've witnessed firsthand: the conflict between scientific understanding and deeply ingrained ideological fear. History has shown us how often human progress falters not because of lack of knowledge, but because of our refusal to confront it. Extremism thrives when we're afraid of what new truths imply about our place in the world.

That's why I write stories like this—not to provide answers, but to offer a space where uncertainty is allowed, even encouraged. Where facts aren't enemies of meaning. Where evidence can be the beginning of wonder, not its end. Where we don't fear questioning our place in nature, but embrace it as one of the most human things we can do.

Thank you for reading. Thank you for thinking. And thank you for letting these strange digital echoes speak to you.

[About the Author]

Gabe Stroup is a graduate student, aspiring scientist, and science communicator, with a deep passion for exploring the intersection of nature and humanity. When not immersed in scientific research, Gabe writes fiction that illustrates challenging ideas about nature, the use of technology, and the future of scientific thought.

Gabe's *Ghosts of the Biosphere* series was born out of a lifelong interest in biology, history, and scientific progress, and from a fascination with how digital ecosystems could live on as a natural continuation when it becomes extinct in the real world, and what those developments could mean for our understanding of life itself. He was also inspired from other sci-fi writers like Michael Crichton and Ted Chiang.

This trilogy is Gabe's first foray into sci-fi novel writing, combining mystery, ethics, and environmental science in a narrative that asks readers to question the role of humans as stewards of both the real and the digital biospheres we create. He has previously written short story collections and epic stories that all have something to do with science, history, and human legacies.

When not writing, Gabe can be found exploring the outdoors, delving into complex scientific topics, or pursuing other artistic expressions like art and music. To follow Gabe's work and connect directly, visit: Bit.ly/GabeStroupBooks

Or scan the QR code below.

www.ingramcontent.com/pod-product-compliance
Lightning Source LLC
LaVergne TN
LVHW100301210926
848480LV00003B/983

* 9 7 9 8 2 1 8 6 9 0 1 5 1 *